Table of Contents

INTRODUCTION 5
ARITHMETIC 5
   SETS OF NUMBERS 6
      PRACTICE 1 11
   OPERATIONS WITH INTEGERS 12
      PRACTICE 2 15
   USING THE DECIMAL SYSTEM 16
      PRACTICE 3 18
   ROUNDING 18
      PRACTICE 4 20
   OPERATIONS WITH DECIMALS 20
      PRACTICE 5 21
   DIVISIBILITY RULES 22
   PRIME NUMBERS 23
   FINDING THE LEAST COMMON MULTIPLE AND GREATEST COMMON FACTOR 25
      GRAPHING CALCULATOR HELP: LCM AND GCF 27
      PRACTICE 6 29
   OPERATIONS WITH FRACTIONS 30
      PRACTICE 7 31
   ORDER OF OPERATIONS 33
      PRACTICE 8 36
      PRACTICE 9 38
      GRAPHING CALCULATOR HELP: ORDER OF OPERATIONS 40
   CONVERTING FRACTIONS, DECIMALS, AND PERCENTS 41
      GRAPHING CALCULATOR HELP: DECIMALS AND FRACTIONS 44
      PRACTICE 10 45
   RATIOS, RATES, AND PROPORTIONS 46
      PRACTICE 11 48
   METRIC SYSTEM 49
      PRACTICE 12 51
   SOLVING PERCENT PROBLEMS 52

PRACTICE 13 53
PERCENT OF CHANGE 53
PRACTICE 14 54
DISCOUNTS, MARKUPS, TAXES 55
Discounts 55
Markups and Taxes 56
Combining Discount with Markup/Tax 56
PRACTICE 15 57
SIMPLE INTEREST 59
PRACTICE 16 60
SOLVING PROPORTION WORD PROBLEMS 61
PRACTICE 17 63
GEOMETRY 65
TYPES OF ANGLES 66
PRACTICE 18 71
PARALLEL LINES AND ANGLE RELATIONSHIPS 73
PRACTICE 19 75
TRIANGLES 76
PRACTICE 20 78
FINDING MISSING ANGLES IN A TRIANGLE 80
PRACTICE 21 82
CLASSIFYING QUADRILATERALS 84
PRACTICE 22 87
CLASSIFYING POLGONS 88
PRACTICE 23 89
CLASSIFYING SOLIDS 90
PRACTICE 24 91
INTRODUCING COORDINATE GEOMETRY 93
PRACTICE 25 94
FINDING THE MIDPOINT 95
PRACTICE 26 96
PYTHAGOREAN THEOREM AND FINDING SIDES OF A RIGHT TRIANGLE 97
PRACTICE 27 99

FINDING THE DISTANCE BETWEEN TWO POINTS 101
   PRACTICE 28 102
FINDING PERIMETER, AREA AND CIRCUMFERENCE 104
   PRACTICE 29 108
FINDING AREA AND PERIMETER OF COMPOSITE SHAPES 111
   PRACTICE 30 113
CONGRUENT TRIANGLES 114
   PRACTICE 31 117
SIMILAR TRIANGLES 119
   PRACTICE 32 121
FINDING VOLUME 123
   PRACTICE 33 124

ALGEBRA 125
  SIMPLIFYING EXPRESSIONS 125
   PRACTICE 34 128
  EVALUATING EXPRESSIONS AND USING ORDER OF OPERATIONS 129
   PRACTICE 35 130
  SOLVING ONE-STEP AND TWO-STEP EQUATIONS 130
   PRACTICE 36 133
  SOLVING MULTISTEP EQUATIONS 134
   PRACTICE 37 135
  SOLVING INEQUALITIES 138
   PRACTICE 38 140
  FINDING SLOPE AND RATE OF CHANGE 143
  FINDING SLOPE FROM A GRAPH 144
  FINDING SLOPE FROM TWO POINTS 146
   PRACTICE 39 146
  LINEAR EQUATIONS 148
  GRAPHING LINEAR EQUATIONS 148
   GRAPHING CALCULATOR HELP: GRAPHING LINES 151
   SPECIAL CASES 151
   PRACTICE 40 152
  SOLVING SYSTEMS OF EQUATIONS 155

CALCULATOR HELP: SYSTEMS	157
PRACTICE 41	158
STATISTICS AND PROBABILITY	161
ANALYZING GRAPHS	161
PRACTICE 42	162
DESCRIPTIVE STATS	163
CONSTRUCTING A BOXPLOT	163
CALCULATOR HELP: STATISTICS	165
PRACTICE 43	166
PROBABILITY	167
PRACTICE 44	169
SOLUTIONS	170
APPENDIX I: TABLES	183
Table 1: Calculator Quick Reference	183
Table 2: Venn Diagram of Real Numbers	185
Table 3: Decimal System	186
Table 4: Large Number Names	186
Table 5: Divisibility Rules	187
Table 6: Table of Prime Numbers	188
Table 7: Common Percent, Decimal, Fraction Conversions	189
Table 8: Metric System	190
Table 9: Geometry Terms	191
Table 10: Angles	192
Table 11: Triangles	193
Table 12: Quadrilaterals	194
Table 13: Polygons	195
Table 14: Solids	196
Table 15: Algebraic Properties	197
APPENDIX II: FORMULAS	198
REFERENCES	201
GRAPH PAPER	203

INTRODUCTION

The goal of this book is to provide a basic understanding of mathematics at an intro to college level. The book is designed to go along with a course of Intro to College Math for those pursuing Nursing AAS or similar programs. It is also designed as a refresher for adult students going back into the math classroom. The course is divided into four main sections: Arithmetic, Geometry, Algebra, and Statistics/Probability. This covers the basics. Adding, subtracting, multiplying, and dividing with decimals and fractions. It also discusses problems with percents, fractions, and proportions in algebraic and real-world contexts. Then we delve into some basic geometry problems. We will look at the basics of what is usually covered in an Algebra I course in high school: substitution, solving equations, solving inequalities, graphing lines, solving systems. And finally, end with how to interpret data and graphs with some descriptive statistics and probability. This book is an expanded form of my lecture notes and includes extra explanations, examples, and practice. If you get stuck with the practice or just want to check your answers, then check with BOB. Solutions to practice sets are in the Back Of the Book. Throughout the book you will also find GRAPHING CALCULATOR HELP sections which will guide you through using a TI-84 series graphing calculator.

ARITHMETIC

Arithmetic is the branch of mathematics that most of us start with, studying numbers, their properties, and operations with those numbers. Here in this section we will be examining sets of numbers, the decimal system, order of operations, fractions, decimals, percents, proportions, and applications of these concepts.

SETS OF NUMBERS

What is a number? How many different types of numbers are there? Imagine going back in time to when numbers were first "invented." What do you think they were used for at first? As we go through this imaginary journey, we will come across different **sets**, or groups, of numbers.

The first set that we come across is the **NATURAL** numbers. These are the counting numbers: 1, 2, 3, 4, … Symbolized by $\mathbb{N}$. If all we care about is counting and adding things up, this is all we'd need, but there's much more. Imagine cavemen and cavewomen counting sticks and stones[1].

The next set is the **WHOLE** numbers. These include the natural numbers and the special number zero. 0 looks like a donut and a donut has a hole (whole number). 0, 1, 2, 3, 4, … Symbolized by W. In modern times, we tend to take zero for granted, but not all ancient civilizations had a concept of zero. It was invented independently in Mesopotamia, Mayan empire, and India starting around 3 BC[2]. Besides signifying nothing, zero is important as a place holder. Without zero, we would end up like the Romans who had to have separate symbols for 1 = I, 10 = X, 100 = C, and 1000 = M. A simple number like 1999 becomes MCMXCIX.

Then we start losing things, going into debt, and we need to keep track of loss. To do this we use the **INTEGERS**. The integers include the whole numbers and their opposites. Keep in mind that the opposite of 0 is still 0; it's neutral. These integers can be listed, but they go on forever, forward and backward. …, −3, −2, −1, 0, 1, 2, 3, … Symbolized by $\mathbb{Z}$.

---

[1] (Maxfield, 2009) (Number Systems, n.d.)
[2] (Matson, n.d.)

---

[3] Picture courtesy of Tomas Quinones under Creative Commons License CC BY-SA 2.0 (Quinones, 2012)

Why use a Z for integers, when it clearly starts with I? The Z stands for the German word *Zahlen*, meaning numbers. Can you imagine how negative numbers were invented? Negative numbers were first used around the 7th century AD in India to keep track of debts[4].

The integers are fine for most things, we can count, add, subtract, multiply, but once we start dividing, we run into trouble. That's where the **RATIONAL** numbers come in. Rational comes from the root *ratio,* which means to cut or divide. If we have one pizza and cut it into eight parts, then we eat one slice that's $\frac{1}{8}$ of the pizza. If we eat two slices, we have $\frac{2}{8}$ but that's the same as $\frac{1}{4}$. Rational numbers include fractions that can be written using the integers. We can also write these as decimals, but the decimals either stop like $\frac{1}{4} = .25$ or they repeat themselves like $\frac{2}{3} = .666\ldots$ which we can also write as $.\overline{6}$, the bar showing which digits repeat. For example, $.12\overline{34} = .12343434\ldots$ Rationals include decimals, fractions, and percents and you will learn how to convert among each of them. Rationals are symbolize by $\mathbb{Q}$. The Q here stands for quotient.

Since a rational is any number that can be written as a fraction using integers, the set of *rational numbers also includes the integers*, because any integer can be written as a fraction over 1. For example, $-2 = -\frac{2}{1}$. Writing an integer this way can be helpful at times, but remember that when you give solutions, you should always **reduce fractions** unless otherwise stated. So far, the sets of numbers are like Russian nesting dolls. The natural numbers are the

---

[4] (Rogers, 2008)

"smallest" set which is inside the whole numbers. The whole numbers are inside the integers. And finally, the integers are inside the "largest" set of the rationals.

There is another set of numbers that is slightly crazy, the **IRRATIONAL** numbers. In our imaginary time travel, rational numbers and irrational numbers were probably invented in ancient Greece, usually attributed to Pythagoras[5]. Irrational numbers cannot be written as fractions using integers. When we write an irrational number as a decimal, we get a never-ending string of non-repeating digits. The most famous of these is probably π, which is approximately 3.14 or $\frac{22}{7}$, but looks more like 3.14159265358 ... and goes on and on like a pink bunny, never repeating itself[6]. Euler's number is another example, $e \approx 2.718281828459$ ... Another interesting example is the number .12345678910111213... which has a pattern but notice it will never repeat and never end. Other examples of irrational numbers include square roots and cube roots which don't come out evenly, such as $\sqrt{2}$, $\sqrt{5}$ or $\sqrt[3]{9}$. However, not all roots are irrational; $\sqrt{16} = 4$ so it's perfectly rational (as well as integer, whole, and natural). There is no standard abbreviation for irrationals. I usually just write *Irrat*. to abbreviate the irrational numbers.

All these types of numbers (Natural, Whole, Integers, Rational, and Irrational) together comprise what's called a **REAL NUMBER**, abbreviated by $\mathbb{R}$. Throughout this course, every number that we use will be a real number. There are more types and sets of numbers in the mathematical universe beyond the scope of this course. You may remember imaginary and

---

[5] Some scholars believe Hippasus of Metapontum discovered irrationality of $\sqrt{2}$ (Wikipedia Contributors, 2019a)
[6] (Bailey, Borwein, Borwein, & Plouffe, 1997)

complex numbers from high school algebra, but there's even more beyond that[7]. A number is just a mathematical object used for quantifying and manipulating, through certain operations, something we want to describe.

So again, we are going to be keeping it Real throughout this book. We must keep in mind the nesting of the naturals inside the whole numbers, the whole numbers inside the rationals, and the rational numbers inside the reals. If you're not rational, you're irrational, but you're still real. Here's a Venn Diagram showing the relationship between the sets of numbers.

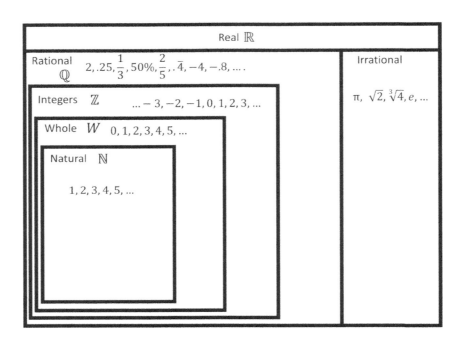

You will be able to determine all the set or sets to which a given number belongs.

Ex. 1 What sets of numbers does 2 belong to?

2 is a natural number, whole number, integer, rational, and real or N, W, Z, Q, R.

---

[7] hypercomplex, surreal, quaternions, octonions, sedonions, infinitesimal, transfinite and more (Wikipedia Contributors, 2019b)

Ex. 2 What sets of numbers does $\sqrt{2}$ belong to?

Since $\sqrt{2}$ cannot be simplified it is irrational and real or I, R.

Ex. 3 What sets of numbers does $\frac{\sqrt{4}}{3}$ belong to?

In this case $\frac{\sqrt{4}}{3}$ can be simplified to $\frac{2}{3}$ so it can be written as a fraction which makes it rational and real or Q, R.

PRACTICE 1

Name the set or sets to which each number belongs.

1) 6

2) −6

3) $\frac{4}{5}$

4) $\sqrt{53}$

5) 0

6) 3

7) $\frac{20}{13}$

8) π

9) 9

10) −1

11) $\frac{13}{4}$

12) $\sqrt{39}$

13) 6

14) $\sqrt{64}$

15) $\sqrt{61}$

16) $\dfrac{22}{14}$

17) −8

18) 10

19) 0

20) $\sqrt{0}$

OPERATIONS WITH INTEGERS

Let's start with the basics. How do we add, subtract, multiply, and divide integers which could be either positive or negative? Sometimes it may seem difficult to keep track of the signs. Subtraction can always be written as adding the opposite, but let's explain a little further. Let's first think what happens on a number line.

How would you add two positives? 2 + 3. Well you would start at two and since we are adding we would face to the right and then move forward 3 spaces ending up at 5.

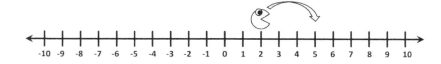

How would you subtract two positives? 2 – 3. Well you would start at two and since we are subtracting, we would face backwards to the left and then move forward 3 spaces ending up at -1.

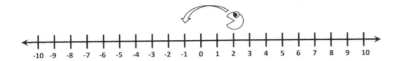

How would you add a positive and a negative? 2 + -3. Well you would start at two and since we are adding we would face forward to the right and then move backwards 3 spaces ending up at -1.

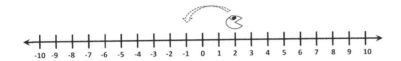

Lastly, how would you subtract a negative from a positive? 2 – -3. Well you would start at two and since we are subtracting, we would face backwards to the left and then move backwards 3 spaces ending up at 5. Remember subtracting we can write as adding the opposite.

Here's another analogy using a basket floating in the sky with weights attached[8]. Imagine that balloons are positive numbers and weights are negative numbers. Also imagine that right now the balloon is completely balance and just floating in the air. If we add balloons which way would the basket go. Of course, it would go up. Adding a positive takes us into the positive "up" direction. What happens when we add a weight, when we add a negative? The basket would go down. Adding a negative takes us in the negative or "down" direction. Then suppose we remove a balloon, we are subtracting a positive, the basket would start to sink down. Subtracting a positive takes us into the negative direction. Lastly imagine removing one of the weights. When we subtract a negative, it's like taking a weight away and our basket would float higher into the positive "up" direction.

Since multiplication is repeated addition and division is the inverse of multiplication, multiplying and dividing must be consistent with the rules we have for addition and subtraction. To interpret a negative multiplied by a negative, imagine you have a direct withdrawal from your checking account for $100. What would happen if that was reversed or someone else paid it for you for the next 12 months? You would have a net effect of positive $1200. That is $-12 \times -100 \; or \; (-12)(-100) = 1200$. Or even simpler, what is the opposite of a negative? It must be a positive. $-(-5) = -1(-5) = 5$. A negative times a negative is a positive. The

---

[8] This analogy was found on Math is Fun (Pierce, 2018)

enemy of my enemy is my friend. Since division is the reverse[9] of multiplication, division must work the same way.

Here's a visual summary of all the properties.

⊕ + ⊕ = ⊕   ⊕ × ⊕ = ⊕
⊖ + ⊖ = ⊖   ⊖ × ⊖ = ⊕
⊕ + ⊖ = ⊕   ⊖ × ⊕ = ⊖
⊕ + ⊖ = ⊖   ⊕ × ⊖ = ⊖

Subtraction rewrite as adding the opposite

Division follows the same signs as multiplication

PRACTICE 2
Simplify the expressions

1) $21 + 24$

2) $(-7) - 2$

3) $7 + (-17)$

4) $10 + (-5)$

5) $(-11)(-10)$

6) $(-8)(-12)$

7) $-10 \div -5$

8) $36 \div -9$

9) $-16 \div 8$

10) $-100 \div 10$

11) $(-6) + (-22)$

12) $2 - 22$

13) $23 + 20$

14) $(-13) - (-24)$

15) $(-11)(14)$

16) $(-3)(10)$

---

[9] inverse

17) $\dfrac{-22}{2}$

18) $\dfrac{-98}{-14}$

19) $(-23)-(-23)$

20) $3-(-18)$

21) $(-23)+(-19)$

22) $11+(-3)$

23) $11-(-7)$

24) $(-24)-5$

USING THE DECIMAL SYSTEM

We use a base 10 or **decimal system** with 10 different digits 0 through 9. The first place value is called the units or ones place. Each place value to the left is ten times as much. We have the tens, hundreds, thousands, then ten thousands, hundred thousands, until we get the millions. It repeats with ten million, hundred million until we get the next big name: billion. When we move to the right of the decimal place each value gets divided by ten. To the right of the decimal we have tenths, hundredths, thousandths, ten thousandths, etc. Notice that the decimal places to the right follow the same pattern and each end in -th. Also, there is no oneths place; it just would not make any sense[10].

---

[10] A "oneth" would have to be $\dfrac{1}{1}$ which is equal to 1

Base 10 (decimal) system

Examples	million	hundred thousand	ten thousand	thousand	hundred	ten	unit	decimal point	tenth	hundredth	thousandth	ten thousandth	hundred thousandth
1 million	1	0	0	0	0	0	0	•					
400 thousands		4	0	0	0	0	0	•					
3 thousand 2 hundred				3	2	0	0	•					
6 hundred seventeen					6	1	7	•					
Twenty						2	0	•					
5 tenths							0	•	5				
8 hundredths							0	•	0	8			
73 hundred thousandths							0	•	0	0	0	7	3

There are other number base systems used in math such as binary (base 2) using 0's and 1's in computer systems. Historically, base 60 was developed by Sumerian culture, where we see the remnants still used in telling time and measuring angles: 60 seconds in a minute, 60 minutes in an hour and 360 degrees in a circle. But our focus in this course will be the "handy" decimal system, as easy to use as counting on your fingers[11]. If we started counting with our fingers and our toes, we may have decided that base 20 was more convenient.

---

[11] (Maxfield, 2009)

PRACTICE 3

Identify the place value of each underlined digit

1. 26, 3<u>4</u>1, 798.1467
2. 67, 153.95<u>8</u>2
3. 38.<u>0</u>19528
4. 0.29<u>8</u>3
5. 1, 5<u>8</u>2, 949, 028.93
6. 3.1<u>4</u>15926535897
7. 45.9<u>8</u>7
8. 4, 21<u>6</u>.3893
9. 8932.<u>6</u>832
10. <u>2</u>, 485, 692.13
11. 2, 4<u>8</u>5, 692.13
12. 3.141<u>5</u>926535897
13. 513.89<u>9</u>731
14. 513.8<u>9</u>9731
15. 51<u>3</u>.899731
16. 26, 341, 7<u>9</u>8.1467
17. 26, 341, 79<u>8</u>.1467
18. <u>3</u>.1415926535897
19. 3.1415<u>9</u>26535897
20. 2<u>6</u>, 341, 798.1467

ROUNDING

When we round to a certain place value, we want to see which multiple of that place value our given number is closest to. For example, say you need to pay $263.18, but all you have is $100 bills. So, if we count by 100's, we would have 100, 200, 300 ... 263.18 falls between 200 and 300, but which is it closest to? 263.18 is closer to 300 so it rounds to 300.

When you go to buy gas, you may notice sometimes the price is something like $2.459. The price of gas may have a number in the thousandths place which is known as a mill[12]. So, if

---

[12] A mill is one thousandth of a dollar (Wikipedia Contributors, 2019e)

you buy 12 gallons of gas it would we get 12(2.459) = 29.508 which rounded to the nearest cent (hundredth) we get $29.51.

When rounding look at the place value you want to round to. I suggest underlining the digit in that place value. Then look at the digit immediately to the right. If it is 4 or less, then we round down. The underlined digit stays the same, all digits to the left will remain the same, and all digits after the underlined digit will become zeros. If it is 5 or more, then we round up. The underlined digit goes up by 1 (if the underlined digit is a 9 then it becomes a 0 and we need to carry a 1 to the place value to the left), all digits to the left will remain the same, and all digits after the underlined digit will become zeros.

Ex. 1 Round 93,614.82 to the hundreds place

93,614.82 the 6 is in the hundreds place

93,614.82 the digit to the right is 1

93,600 the 6 stays a 6, everything after becomes 0. We can drop the trailing zeros after the decimal points. So, 93,614.82 rounded to the hundreds place becomes 93,600.

Ex. 2 Round 93,614.82 to the thousands place

93,614.82 the 3 is in the thousands place. 93,614.82 the digit to the right is 6

94,000 the 3 rounds up to 4, everything after becomes 0. We can drop the trailing zeros after the decimal points. So, 93,614.82 rounded to the thousands place becomes 94,000.

Ex. 3 Round 43,984.371 to the hundreds place

43,9̲84.371 the 9 is in the hundreds place

43,9̲84.371 the digit to the right is 8 so we need to round up

43,9̲84.371 the 9 rounds up to 10 but we need to carry the 1 to the left

44,0̲00 the 3 to the left becomes 4 because we carried the 1 and everything to the right becomes 0. So, 43,984.371 rounded to the hundreds place becomes 44, 000.

PRACTICE 4

Round to indicated place value

1. 96, 399, 533.49 ten thousands
2. 64.249356 hundredths
3. 59, 345.9526 hundreds
4. 13, 590.00245 ten thousandths
5. 1.94566 tenths
6. 45, 938.48 ones
7. 54, 024.24935 thousandths
8. 94, 432.04306 tens
9. 3.1415926535 hundredths
10. 3.1415926535 thousandths
11. 3, 896.13 tens
12. 3, 896.13 hundreds
13. 3, 896.13 thousands
14. 3.1415926535 tenths
15. 3.1415926535 ones
16. 54, 024.24935 thousands
17. 96, 399, 533.49 thousands
18. 96, 399, 533.49 tens
19. 798.54 ones
20. 798.54 tens

OPERATIONS WITH DECIMALS

When adding or subtracting with decimals, it is easiest to rewrite the problem vertically and line up the decimal places. Start adding on the right side and remember to carry when you need to.

$$4.362 + 12.679$$

$$+\begin{array}{r}4.362\\12.679\\\hline17.041\end{array}$$

When multiplying, we don't need to line up the decimal. Just multiply as usual. Then add up the total decimal places in the problem and make sure your answer has the same number of decimal places.

$$\begin{array}{r} 4.2 \\ \times\ .17 \\ \hline 294 \\ 42\ \ \\ \hline .714 \end{array}$$

4.2   1 decimal place
× .17   2 decimal places
.714   3 decimal places

Dividing with decimals is much easier by setting up a long division box. If there is a decimal in the divisor, then move the decimal to the right until it becomes a whole number. Move the decimal in the dividend the same number of spaces. Divide as usual. Continue dividing until you get a remainder of zero or it starts repeating. You cannot use a remainder with decimals. Then put the decimal point directly above the decimal point in the dividend, the number in the box after its been moved.

$$\frac{3.525}{1.5} \Rightarrow 1.5\overline{)3.525} \Rightarrow 15\overline{)35.25}$$

Result: 2.35

PRACTICE 5

Find each sum.

1) $0.8 + 5.1$

2) $4.8 + 1.284$

3) $2.8 + 5.72$

4) $0.1 + 0.483$

5) $3.6 + 4.73$

6) $7.236 + 5.4$

Find each difference.

7) $2 - 0.1$

8) $5.2 - 4.8$

9) $4.1 - 2.42$

10) $4.4 - 0.968$

11) $6.3 - 4.685$

12) $7.6 - 7.186$

Find each product.

13) 1.4×5.42

14) 2.6×1.3

15) 0.5×3.7

16) 9.4×5

17) 8×3.76

18) 9.3×5.7

Find each quotient.

19) $\dfrac{7.6}{0.5}$

20) $\dfrac{9.3}{0.6}$

21) $\dfrac{1.05}{0.3}$

22) $\dfrac{1.6}{3.2}$

23) $\dfrac{3.9}{0.4}$

24) $\dfrac{2.8}{0.56}$

DIVISIBILITY RULES

Pick a random 4-digit number[13], let's say 7978. Well it should be easy to see it's divisible by 2 since it's easy. It's not divisible by 3, 4, 5, 6, 7, 8, 9, 10, 11, 12, 13, ... In fact, the only other factor of 7978 is 3989. Divisibility rules[14] give us shortcuts to determine what a number can be divided by without any remainder. Using the divisibility rules will make it easier to simplify and work with fractions. The table on the next page has divisibility rules for numbers up to 12. Table 4 in the appendix has divisibility rules for numbers up to 19. It's not necessary to remember all these rules. You may already know the rules for 2, 5, and 10. You may find it

---

[13] (NumberGenerator.org, n.d.)
[14] (Wikipedia Contributors, 2019)

helpful to learn the rules for 3, 6 and 9 as well since those will make mental calculations much easier.

	A number is divisible by…	Divisible	Not Divisible
2	The last digit is even (0,2,4,6,8)	2018	2019
3	The sum of the digits is divisible by 3	2019 2+0+1+9=12	2020
4	The last 2 digits are divisible by 4	2020	2021
5	The last digit is 0 or 5	2020	2019
6	Is even and is divisible by 3 (follows 2 rule and 3 rule)	2016 2+0+1+6=9	2019
7	Double the last digit and subtract it from a number made by the other digits. The result must be divisible by 7. (This rule is most difficult on this list)	2016 201-12=189 18-18=0	2017
8	The last three digits are divisible by 8	2024	2020
9	The sum of the digits is divisible by 9	2025 2+0+2+5=9	2019
10	The number ends in 0	2020	2021
11	Add and subtract digits in an alternating pattern (add digit, subtract next digit, add next digit, etc.). Then check if that answer is divisible by 11.	2035 2-0+3-5=0	2037
12	The number is divisible by both 3 and 4 (follows the 3 rule and 4 rule)	2028 2+0+2+8=12	2037 Divisible by 3 but not 4

Now that you now the shortcuts can you figure out what 840 is divisible by? Here's a hint it divisible by all but 2 of the numbers above.

PRIME NUMBERS

A prime number is a natural number that has exactly two factors. Factors are numbers that divide evenly into a given number. For example, 2 is a prime number since the only numbers that divide into 2 are 2 and 1. A natural number that has more than two factors is called composite. Four is a composite number because it is divisible by 1, 2, 4. The number one is an unusual number that is neither prime nor composite. Primes are considered the building blocks of numbers[15]. The first few primes are: 2, 3, 5, 7, 11, 13, 17, 19, 23, 29, 31… The list of primes is infinite, so there is no largest prime[16]. Every number can be factored uniquely into its prime factors[17]. We will be using primes mostly to simplify fractions but also find all the factors of a number, the greatest common factor and the least common multiple.

<u>Ex. 1</u> Find the prime factorization and then all the factors of 840

We can make a factor tree to find the prime factorization

```
    840
   /  \
  2   420
      /  \
     2   210
         /  \
        2   105
            /  \
           3   35
               / \
              5   7
```

Making the factor tree will also help to find all the factors. When making a list of all factors it helps to list them in pairs. Of course, we start with 1 and itself 840. We can see from the tree that 2 is a factor and its other factor is 420. 3 also divides into 840 giving 280. Continue through the rest of the numbers. Of course, we don't need to check 9 because we only have a single 3 in our factor tree. Likewise, we don't have to check any primes not in the factor tree. Once we get to the middle pair in our list 28 and 30, we know we are done. We don't have to check any other numbers. There can't be a factor higher because it would have to be paired with a lower number and we checked through all the lower numbers. The complete list is shown here:

1, 2, 3, 4, 5, 6, 7, 8, 10, 12, 14, 15, 20, 21, 24, 28, 30, 35, 40, 42, 56, 60, 70, 84, 105, 120, 140, 168, 210, 280, 420, 840

---

[15] (Caldwell, 1996)
[16] (Alfed, 1996)
[17] This is called the Fundamental Theorem of Arithmetic (Smith, 2015)

FINDING THE LEAST COMMON MULTIPLE AND GREATEST COMMON FACTOR

A multiple is when we count by a number, it is divisible by the given number without any remainder. Multiples of 2 are 2, 4, 6, 8, 10, 12, 14, 16, 18, 20, 22, 24 … Multiples of 3 are 3, 6, 9, 12, 15, 18, 21, 24 … The lists are infinite because we can keep counting as high as we want. As you can see some of the numbers are the same on both list such as 6, 12, 18 and 24 are all common multiples. There would be an infinite number of matches if we kept writing both lists. The least common multiple (LCM) is the smallest match which for 2 and 3 would be 6.

As stated previously, a factor is a number that divides into another number. The factors of 24 are 1, 2, 3, 4, 6, 8, 12, 24. The factors of 60 are 1, 2, 3, 4, 5, 6, 10, 12, 15, 20, 30, 60. The number of factors is limited. Again, there are several numbers in both lists: 1, 2, 3, 4, 6, and 12. The greatest common factor (GCF) is the largest match which in this case would be 12.

Usually when we find LCM and GCF, we make lists as we did above. However, there is a method to find both at the same time which is especially easier for larger numbers. We will factor both numbers completely and place the numbers in a Venn Diagram[18]. A The product of all the numbers in the intersection (the overlap of the two circles) will be the GCF. While the product of all the numbers overall, the union, will be the LCM. This method also shows the close relationship between the least common multiple and greatest common factor in a more mathematically beautiful way.

---

[18] A Venn Diagram shows the relationship between members of different sets. Usually 2 or 3 intersecting circles are used. (Wikipedia Contriutors, 2019d)

Ex. 1 Find the LCM and GCF of 48 and 60

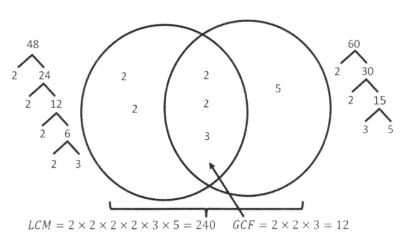

$LCM = 2 \times 2 \times 2 \times 2 \times 3 \times 5 = 240 \quad GCF = 2 \times 2 \times 3 = 12$

We draw a Venn Diagram. The left circle will be for 48 while the right circle will be for 60. We start by making a factor tree for each number. They have 2, 2, and 3 in common. The numbers they have in common go in the overlap, intersection, of the two circles. 48 has two extra factors of 2, so we write those in the left circle. While 60 has an extra factor of 5 and we write that in the right circle. We get the GCF by multiplying the three numbers in the intersection which comes out to 12 and the LCM is the product of all the factors in the Venn diagram, coming out to 240.

Ex. 2 Find the LCM and GCF of 52 and 35

Again, we start by doing a factor tree for each number. They have no common prime

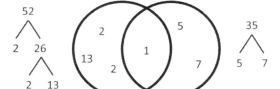

factors in their trees. But there still is a common factor. We will write 1 in the intersection since there is no other factors they have in common. Whenever you use this method and there doesn't seem to be a factor to write in a section, you can always put 1 in there. The GCF is the intersection which is just 1 in this case. The LCM is the product $2 \cdot 2 \cdot 5 \cdot 7 \cdot 13 = 1820$.

Ex. 3 Find the LCM and GCF of 40, 24, and 84

This time we draw a Venn Diagram with three circles. Do a factor tree for each of the numbers.

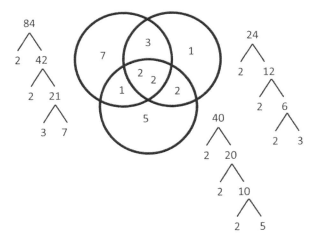

Then start with the innermost section, the intersection of all three circles. What do they all have in common? They have two 2's in common. Then fill in the other intersections between two circles. 24 and 40 have another 2 in common. 24 and 84 have a 3 in common.

And 84 and 40 have no other factors in common, besides 1. Write any remaining factors in. The GCF is the intersection of all three. GCF(40, 24, 84) = 4. The LCM is the product of all the numbers in the circles. LCM(40, 24, 84) = 840.

GRAPHING CALCULATOR HELP: LCM AND GCF

You will have to be able to find the LCM and GCF by hand, but this section will show you how to find the LCM and GCF[19] using the calculator functions. This will mainly be used for checking your answers.

EX. 1 Find the LCM and GCF of 54 and 12

Press the MATH button then go over to the NUM and down to 8:lcm(. Press ENTER. Type 54 then , then 12. The comma key is right above the 7 key. Then press ENTER. Finding the GCF

---

[19] The least common multiple, LCM, is also the same as least common denominator, LCD. The greatest common factor, GCF, is also the same as greatest common denominator, GCD.

is similar. Press the [MATH] button then go over to the NUM and down to 9:gcd(. Type gcd(54,12). Then press [ENTER].

EX. 2 Find the LCM and GCF of 36, 24, 30

The calculator is limited in that it cannot find the LCM and GCF of three numbers directly. But there is a work around. For the LCM of three numbers, first find the LCM of the first two numbers, then take the LCM of that answer with the third number. LCM(36,24)=72. LCM(72, 30)=360. The LCM(36, 24, 30) = 360. Similar for the GCF. Find the GCF (GCD) of the first two numbers, then take the GCD of that answer with the third number. GCD(36,24)=12. GCD(12, 30)=6.

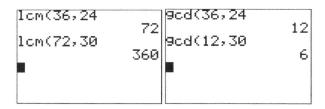

PRACTICE 6

Find LCM and GCF of each pair

1. 40, 24
2. 21, 44
3. 28, 20
4. 60, 35
5. 48, 30
6. 45, 72
7. 56, 42
8. 45, 10
9. 32, 36
10. 21, 42

11. 56, 48
12. 16, 36
13. 48, 16
14. 20, 46
15. 40, 24, 16
16. 18, 12, 30
17. 45, 75, 30
18. 7, 8, 9
19. 56, 20, 40
20. 27, 16, 32

OPERATIONS WITH FRACTIONS

When adding or subtracting fractions we need a common denominator (the bottom of a fraction). When the denominator is the same then we add (or subtract) the numerators and the denominator will stay the same. We find a Least Common Denominator by finding the Least Common Multiple of the denominators. Make sure your final answer is a reduced by dividing the numerator and denominator by their greatest common factor.

Ex 1 $\frac{3}{10} + \frac{1}{10} = \frac{4}{10} = \frac{2}{5}$ Here, we have common denominators, so we can add right away.

Ex 2 $\frac{5}{6} + \frac{3}{8} = \frac{20}{24} + \frac{9}{24} = \frac{29}{24} \text{ or } 1\frac{5}{24}$ In this example, we need a common denominator for 6 and 8. The LCD is 24. We multiply the first fraction by $\frac{4}{4}$ and we multiply the second fraction by $\frac{3}{3}$.

With subtraction, we still need a common denominator and follow the steps above.

Ex 3 $\frac{5}{6} - \frac{3}{8} = \frac{20}{24} - \frac{9}{24} = \frac{11}{24}$

With mixed numbers it is probably easiest to change both fractions to improper fractions, then find a common denominator as we do below.

Ex 4 $2\frac{1}{2} - 1\frac{2}{3} = \frac{5}{2} - \frac{5}{3} = \frac{15}{6} - \frac{10}{6} = \frac{5}{6}$

When multiplying, it is imperative that we change all fractions to improper fractions. We don't need a common denominator. We can make multiplying easier if we can reduce before multiplying. You can cancel/reduce any numerator (top) with any denominator (bottom). We multiply the numerators to get the new numerator. We multiply the denominator to get the new denominator. Make sure your final answer is simplified.

Ex 1 $\frac{5}{6} \cdot \frac{3}{8} = \frac{5}{2} \cdot \frac{1}{8} = \frac{5}{16}$ In this example we can reduce the 3 and 6 first before multiplying.

Ex 2 $2\frac{5}{8} \cdot 1\frac{1}{3} = \frac{21}{8} \cdot \frac{4}{3} = \frac{7}{2} \cdot \frac{1}{1} = \frac{7}{2}$ Here we change the mixed numbers to improper fractions first. Then reduce the 21 and 3, reduce the 4 and 8. It makes it much easier to reduce first, then multiply across.

When dividing, again it is imperative that we change all fractions to improper fractions. After changing to improper fractions, we change division to multiplying by the reciprocal. This means we change the division to a multiplication and we "flip" the second fraction. Just like multiplication we don't need a common denominator. Follow the steps for multiplication. Finally, make sure your answer is reduced.

PRACTICE 7

Find each sum.

1) $\frac{1}{2} + \frac{3}{2}$

2) $\frac{2}{3} + \frac{1}{2}$

3) $2\frac{3}{5} + 2\frac{1}{2}$

4) $2\frac{1}{6} + 4\frac{1}{2}$

5) $2 + \frac{1}{4}$

6) $2\frac{3}{4} + 2\frac{3}{5}$

Find each difference.

7) $\dfrac{5}{7} - \dfrac{1}{7}$

8) $4\dfrac{1}{3} - 4\dfrac{1}{4}$

9) $1\dfrac{3}{4} - \dfrac{3}{2}$

10) $\dfrac{5}{4} - \dfrac{1}{3}$

11) $2\dfrac{5}{8} - 2\dfrac{1}{6}$

12) $\dfrac{3}{2} - \dfrac{7}{8}$

Find each product.

13) $\dfrac{6}{5} \cdot \dfrac{9}{10}$

14) $\dfrac{3}{2} \cdot \dfrac{4}{5}$

15) $3 \cdot \dfrac{17}{9}$

16) $\dfrac{3}{10} \cdot \dfrac{5}{9}$

17) $3\dfrac{3}{8} \cdot \dfrac{7}{9}$

18) $5\dfrac{1}{3} \cdot \dfrac{7}{4}$

Find each quotient.

19) $\dfrac{5}{9} \div \dfrac{1}{2}$

20) $\dfrac{5}{3} \div \dfrac{7}{6}$

21) $9\dfrac{1}{10} \div 3\dfrac{9}{10}$

22) $1\dfrac{3}{4} \div 8\dfrac{1}{3}$

23) $2 \div \dfrac{7}{6}$

24) $4\dfrac{2}{3} \div 9$

ORDER OF OPERATIONS

In some arithmetic problems there are multiple operations to perform in the same problem. Sometimes on social media, you see viral math problems. For example, this simple looking arithmetic problem had over 5 million views on Mind Your Decisions[20] channel on YouTube: $6 \div 2(1 + 2) = ?$. Depending on the order that you do the operations, you end up with different answers. How can everyone decide the correct way and end up with the same answer? Fortunately, mathematicians have developed a system to determine the order of operations. It can easily be remembered by the mnemonic PEMDAS[21] or Please Excuse My Dear Aunt Sally.

The first step is **P** which stands for parentheses () but also includes any other grouping symbol such as brackets [], braces {}, absolute value | |, and fraction bar – . Resolve what is inside each of these first.

Parentheses
Exponents
Multiplication & **D**ivision
Left to Right
Addition & **S**ubtraction
Left to Right

Absolute Value is the distance from zero[22]. So $|2| = 2$ and $|-2| = 2$, since both 2 and -2 are 2 units away from zero. Absolute value will always be nonnegative (positive unless we take the absolute value of 0 which is 0). The next step **E** is for exponents, such as squaring and cube, but also includes roots, square roots, cube roots, etc. Then comes **MD** which stands for multiplication and division which are done at the same time as we go left to right. Multiplication is sometimes shown by different symbols such as ×,·,∗, by using parentheses

---

[20] (MindYourDecisions, 2016)
[21] In some places, this is also called GEMDAS, BODMAS, BEDMAS, BIDMAS. This is only a change in name and not actual operations (Wikipedia Contributors, 2019c).
[22] This definition of absolute value works for all types of numbers even complex numbers in a plane.

2(3), or when we do algebra, by writing a number in front of a variable 2x. Division can be shown by ÷, but in college algebra we usually use a fraction bar like this $\frac{6}{2}$. Lastly **AS** for addition and subtraction, again as we go left to right.

<u>Ex. 1</u> Simplify $(4 + 2 \cdot 3)|-6 - (-3)|$

Simplify $(4 + 2 \cdot 3)|-6 - (-3)|$ First notice we have two groups of parentheses and one set of absolute value bars.

$(4 + 2 \cdot 3)|-6 - (-3)| = (4 + 6)|-6 - (-3)|$ We look at the first parentheses and we have addition and multiplication (shown by a dot).

$(4 + 6)|-6 - (-3)| = (10)|-6 - (-3)|$ Now we can add. The parentheses around the 10 are holding it and telling us that 10 will get multiplied by what we get from the absolute value part.

$(10)|-6 - (-3)| = (10)|-6 - -3| = (10)|-6 + 3|$ The parentheses around the -3 are just telling us to watch out because there's a negative there. We can remove it. When we subtract, it's the same as adding the opposite. So, subtracting a negative 3 is the same as adding a positive 3.

$(10)|-6 + 3| = (10)|-3|$ We can now add the -6 and 3. Notice we don't do anything with the absolute value bars until we get down to a single number inside the absolute value.

$(10)|-3| = (10)(3) = 30$ Now we take the absolute value of -3 which is 3. The parentheses tell us to multiply 10 and 3 to get our final answer of 30.

Ex. 2 Simplify $\frac{1.5-4.1+(0.5)(-3.8)}{-0.4}$

Simplify $\frac{1.5-4.1+(0.5)(-3.8)}{-0.4}$ Here notice we have a fraction bar, so we need to simplify everything in the numerator and everything in the denominator before resolving the fraction. We start on the top in the numerator. We have subtraction, addition, and multiplication. We start with multiplication shown by the parentheses. Multiplying a positive by a negative, we get a negative. $\frac{1.5-4.1+-1.9}{-0.4}$ Now we have subtraction and addition[23] and we proceed left to right. $\frac{-2.6+-1.9}{-0.4}$ A negative plus another negative drives us further into the negatives. $\frac{-4.5}{-0.4}$ A negative divided by a negative gives us a positive. Our final answer is 11.25.

Ex. 3 Simplify $(\frac{3}{2}-1) \div (-3\frac{1}{2} - \frac{-3}{4})$

In this one we have order of operations with fractions. We have subtraction, parentheses, and division going on. We need to simplify what's in each of the parentheses first. When we subtract, we need common denominators for each parenthesis $(\frac{3}{2}-\frac{2}{2}) \div (-3\frac{2}{4} - \frac{-3}{4})$

Now we can subtract in the first parentheses and we also change the mixed number to an improper fraction in the second parentheses $(\frac{1}{2}) \div (\frac{-14}{4} - \frac{-3}{4})$ Subtracting a negative is the same as adding a positive $\frac{1}{2} \div (\frac{-14}{4} + \frac{3}{4})$ Now we can add easily to get $\frac{1}{2} \div \frac{-11}{4}$ When we divide

---

[23] Adding a negative can also be written as subtraction. $6 + -4$ is the same as $6 - 4$.

with fractions it's the same as multiplying by the reciprocal, so we flip the second fraction $\frac{1}{2}\frac{4}{-11}$

Finally, we cancel out or reduce the 4 and 2 to get our final answer $\frac{1}{1}\frac{2}{-11} = -\frac{2}{11}$

PRACTICE 8

Evaluate each expression.

1) $(-2) - (-3) - (5)(3-5)$

2) $(3 - |5|)(6+6)$

3) $(-4) - 3 + \dfrac{(-4)-6}{5}$

4) $6 - \dfrac{(-11)-1}{3} - 2$

5) $\left(\dfrac{-10}{-2}\right) + 4 - ((-6)-(-2))$

6) $\dfrac{3+3}{(-1)+4-4}$

7) $|5| - |(-1)+3|$

8) $(-5)^2 - (1-2-6)$

9) $\dfrac{(-8)(2)}{(-4)(5-4)}$

10) $(6-3)^3 - ((-5)-5)$

11) $\dfrac{(-5.6) - |5.1-(-1.9)|}{-2.1}$

12) $(-4.6)(-5.4) + |(-1.49)| + 5.5$

13) $\dfrac{(0.2)(-3.5)}{0.7} - |(-2.7)|$

14) $0.5 - (-1.9) + 1.3 - (1.3)(3.2)$

15) $4.8 + (-0.4)^2 - 2.2^2$

16) $-\dfrac{(3.2)(-3.06)}{|2.2 - 1|}$

17) $(-2) - (-3.1) - (5.4)(-2.2) - (-1.3)$

18) $((-5.6) - (3.2 + 1.7))((-1.1) - 2.6)$

19) $\dfrac{(3.3)(4.5)}{-5.4} + (-5.8) + 0.3$

20) $(-3.6)((-4.7) - |(-0.2)|) + 1.3$

21) $1\dfrac{3}{4} - \left(-3\dfrac{1}{3}\right)\left(-\dfrac{3}{2}\right)$

22) $1 - \left(2\dfrac{1}{2} - 2\dfrac{1}{3}\right)$

23) $\left(\left(\dfrac{2}{3}\right)\left(1\dfrac{1}{5}\right)\right)\left(3\dfrac{2}{3}\right)$

24) $\left(-\dfrac{1}{4}\right) - \left(\left(-\dfrac{1}{2}\right) - 2\right)$

25) $(-2)\left(-3\dfrac{5}{6}\right) + \dfrac{2}{3}$

26) $\left(3\dfrac{1}{5}\right)\left(\dfrac{1}{4}\right) - \dfrac{1}{2}$

27) $\left(\dfrac{2}{3} - \left(-2\dfrac{1}{3}\right)\right)\left(-\dfrac{5}{6}\right)$

28) $\left(1\dfrac{1}{3} - (-2)\right)\left(\dfrac{3}{4}\right)$

29) $3\dfrac{5}{6} - \dfrac{2}{3} - \dfrac{3}{4}$

30) $\left(\dfrac{1}{4}\right)\left((-1) - \left(-\dfrac{1}{2}\right)\right)$

PRACTICE 9
Evaluate each using the values given.

1) $np \cdot \dfrac{m^2}{|m|}$; use $m = -4$, $n = -1$, and $p = 3$

2) $|y - x|(x - y)$; use $x = -6$, and $y = -2$

3) $|p - q| + pq$; use $p = 6$, and $q = 3$

4) $z - |-z| + y - z$; use $y = 1$, and $z = 2$

5) $(x)(x^3)(y + y)$; use $x = -1$, and $y = 2$

6) $(z^2 x)(|z|)$; use $x = 2$, and $z = -2$

7) $|x| + (z)(z + x)$; use $x = 5$, and $z = -1$

8) $|y| - (z + y + z)$; use $y = -5$, and $z = 2$

9) $z^2 - (z)(y + x)$; use $x = 4$, $y = -4$, and $z = -6$

10) $\dfrac{|m + n| + n}{m}$; use $m = 4$, and $n = -4$

11) $m(m - |q|)$; use $m = -1.3$, and $q = -3.1$

12) $bc + \dfrac{a}{a}$; use $a = 4.1$, $b = -4.4$, and $c = 2.2$

13) $q + \dfrac{mq}{q}$; use $m = 4.98$, and $q = -4.26$

14) $b - c^2 a$; use $a = -5$, $b = -1.7$, and $c = -3.1$

15) $m - mp$; use $m = 3\dfrac{1}{3}$, and $p = \dfrac{3}{4}$

16) $h - j + k$; use $h = 1\dfrac{3}{5}$, $j = \dfrac{4}{5}$, and $k = 3\dfrac{1}{2}$

17) $z - xy$; use $x = \dfrac{1}{5}$, $y = 2\dfrac{1}{3}$, and $z = 2\dfrac{5}{6}$

18) $(r - p)^2$; use $p = 1$, and $r = \dfrac{3}{2}$

19) $\dfrac{x}{xz} - |y|$; use $x = \dfrac{3}{4}$, $y = -\dfrac{4}{5}$, and $z = -1\dfrac{1}{3}$

20) $\dfrac{hj + j - h}{k}$; use $h = -1\dfrac{3}{5}$, $j = \dfrac{2}{3}$, and $k = -1$

GRAPHING CALCULATOR HELP: ORDER OF OPERATIONS

The calculator can perform all the operations we've discussed so far. The basic operations for addition, subtraction, multiplication and division have their own keys on the right-hand side: [+] [−] [×] [÷]. The calculator makes a strong distinction between subtraction and a negative. When you want to put a negative number in the calculator you must use this key [(−)] located at the bottom of the calculator. When you want to square a number, you can use the [x^2] on the left side. If you square a negative number, you must type it in parentheses to get the correct answer. A negative number squared should give a positive answer. You can also use the caret key [^] to raise a number to any exponent. Square root is another common operation which is easy with the calculator, just press [2ND] [x^2]. To take the cube root of a number, the easiest way is to press the math [MATH] and select $4: \sqrt[3]{(}$. The calculator can also do absolute value which you will also find by pressing the [MATH] button then going over to NUM and selecting ABS. The newer calculators also allow you to type in fractions using [ALPHA] then [Y=] selecting [1: n/d] then press [ENTER].

ARITHMETIC

The calculator can perform order of operations and you can type in the whole problem. Try with $\dfrac{3+2|\sqrt{4}-2^3|}{-5}$

Here's how to type it in:

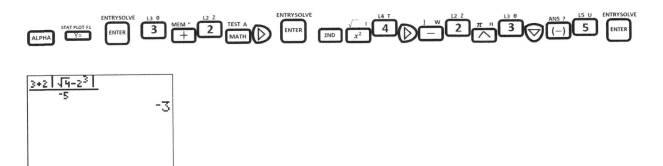

CONVERTING FRACTIONS, DECIMALS, AND PERCENTS

In this section we discuss how to convert among fractions, decimals and percents. To convert a fraction into a decimal, do the division. A fraction bar tells us to divide the numerator (top number) by the denominator (bottom number).

Ex. 1 $\dfrac{3}{5}$ means 3 divided by 5.

```
     0.6
5 | 3.0
    30
    ──
     0
```
By doing the long division, we see 3/5=.6

Ex. 2 $\dfrac{5}{8}$ means 5 divided by 8.

```
     0.625
8 | 5.00
    48
    ──
    20
    16
    ──
    40
    40
    ──
     0
```
By doing the long division, we see 5/8=.625

Ex. 3 $\dfrac{2}{3}$ means 2 divided by 3.

or $.\overline{6}$

```
     0.666...
3 | 2.00
    18
    ──
    20
    18
    ──
    20
    18
    ──
     2
```
By doing the long division, we see 2/3=.666...

To convert a decimal to fraction read the decimal to the correct place value to turn it into a fraction then make sure the fraction is reduced.

Ex. 1   .7 reads as seven tenths which is $\frac{7}{10}$

Ex. 2   .75 reads as seventy-five hundredths which is $\frac{75}{100} = \frac{3}{4}$

Ex. 3   .08 reads as eight hundredths which is $\frac{8}{100} = \frac{2}{25}$

To convert a decimal to percent multiply by 100% (move decimal 2 places to the right and add percent sign).

Ex. 1   .7 becomes 70%

Ex. 2   .001 becomes .1%

Ex. 3   2.5 becomes 250%

To convert a percent to a decimal divide by 100 (move decimal 2 places to the left and drop the percent sign). This is the opposite of changing a decimal to percent shown above.

Ex. 1   30% becomes .3

Ex. 2   4% becomes .04

Ex. 3   108% becomes 1.08

Changing a percent to a decimal is quite easy. Percent % literally means *per cent* or means out of 100. Drop the percent sign and take the number and write it over a denominator of 100. Remember to reduce!

Ex. 1 $60\% = \frac{60}{100} = \frac{3}{5}$

Ex. 2 $120\% = \frac{120}{100} = \frac{6}{5} \text{ or } 1\frac{1}{5}$

Ex. 3 $5\% = \frac{5}{100} = \frac{1}{20}$

To change a fraction into a percent, there are a couple ways we could do it. We could first change the fraction into a decimal then change the decimal to a percent. Or we could take the fraction and multiply it by 100% and make sure to simplify.

Ex. 1 $\frac{1}{4} = \frac{1}{4} \cdot \frac{100\%}{1} = 25\%$

Ex. 2 $\frac{3}{5} = \frac{3}{5} \cdot \frac{100\%}{1} = \frac{3}{1} \cdot \frac{20\%}{1} = 60\%$

Ex. 3 $\frac{2}{3} = \frac{2}{3} \cdot \frac{100\%}{1} = \frac{200\%}{3} = 66\frac{2}{3}\% = 66.\overline{6}\%$

There are some percents, fractions, and decimals that are so common in everyday life that it will be helpful to have some of them memorized for easier, faster conversions. The table on the next page shows some of the more common conversions.

Table of Common Conversions		
Percent	Decimal	Fraction
5%	.05	$\frac{1}{20}$
10%	.1	$\frac{1}{10}$
20%	.2	$\frac{1}{5}$
25%	.25	$\frac{1}{4}$
$33\frac{1}{3}\%$	$.\overline{3}$	$\frac{1}{3}$
50%	.5	$\frac{1}{2}$
$66\frac{2}{3}\%$	$.\overline{6}$	$\frac{2}{3}$
75%	.75	$\frac{3}{4}$
100%	1	1
125%	1.25	$1\frac{1}{4}$
150%	1.5	$1\frac{1}{2}$

GRAPHING CALCULATOR HELP: DECIMALS AND FRACTIONS

Of course, again you should have no problem with converting decimals and fractions by hand, but here's how to use the calculator as a tool to help check your work when you can use one. When you have a decimal in the calculator, you can easily convert to a fraction using

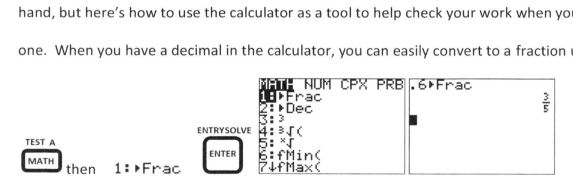

To type in a fraction, you can use the division key ÷ or use ALPHA Y=. If you use the division key and press ENTER the calculator will automatically convert to a decimal.

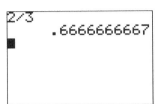

or use MATH 2:▶Dec

PRACTICE 10

Write as a decimal

1. $\frac{4}{5}$
2. $\frac{2}{3}$
3. $\frac{7}{8}$
4. $\frac{5}{2}$
5. $\frac{7}{4}$
6. $\frac{12}{6}$

7. 40%
8. 60%
9. 2%
10. .3%
11. 90%
12. 140%

Write as a fraction

13. $0.\overline{3}$
14. 1.75
15. 0.6
16. 0.25
17. 0.02

18. 2.4
19. 20%
20. 75%
21. 2%

22. 300% 24. 180%

23. 60%

Write as a percent

25. 0.74

26. 2.05

27. 0.30

28. 5

29. 0.35

30. 1.3

31. $\frac{1}{2}$

32. $\frac{6}{5}$

33. $\frac{1}{4}$

34. $\frac{3}{8}$

35. $\frac{2}{5}$

36. $1\frac{1}{2}$

RATIOS, RATES, AND PROPORTIONS

A **ratio** is a comparison of two quantities using the same units. A ratio can be written with a fraction bar, a colon, or using the word *to*. For example, in a classroom there may be 30 women and 20 men. What is the ratio of men to women? $\frac{20}{30}$, 20: 30, 20 *to* 30 or simplified $\frac{2}{3}$, 2: 3, 2 *to* 3. The ratio of women to men is $\frac{30}{20}$, 30: 20, 30 *to* 20 or simplified $\frac{3}{2}$, 3: 2, 3 *to* 2. Notice that the order matters and that when we write a ratio as a fraction, we will leave it as an improper fraction.

Rate is a comparison between two quantities using different units written as a fraction. Example: You travel 580 miles in 8 hours. The rate is $\frac{580\ miles}{8\ hours}$ or simplified $\frac{145\ miles}{2\ hours}$. Most of the time we want the **unit rate**, where the denominator is reduced to one, so we write it as 72.5 mph. When dealing with rates, we need to keep track of the units we are working with.

ARITHMETIC

A **proportion** is an equality of two ratios or two rates. We first identify when a proportion is true, then we will solve a proportion for a missing part. A proportion is true if both sides of the equation are equal when both sides are completely reduced or simplified.

<u>Ex. 1</u> $\frac{8}{14} = \frac{12}{21}$ is a true proportion because we can reduce each side to $\frac{4}{7} = \frac{4}{7}$

<u>Ex. 2</u> $\frac{6}{10} = \frac{26}{40}$ is a false proportion because we can reduce each side to $\frac{3}{5} = \frac{13}{20}$. If we convert to decimals, we would have $.6 = .65$, obviously not true.

Solving proportions can be easy using a method commonly called "cross-multiply and divide." We multiply the numerator (top) of the left by the denominator (bottom) of the right and set it equal to the product of the numerator of right and denominator of the left. Then divide by the **coefficient** of variable, which is the number in front of the variable (being multiplied).

<u>Ex. 1</u> $\frac{3}{8} = \frac{9}{x}$ becomes $3 \cdot x = 9 \cdot 8$ which we write as $3x = 72$. Then divide both sides by 3 to get $x = 24$. Of course, you could see that the numerator on the right is 3 times more than the numerator on the left, so we can multiply the denominator of 8 by 3 to get 24 as well.

<u>Ex. 2</u> $\frac{3}{5} = \frac{x}{4}$ becomes $3 \cdot 4 = 5 \cdot x$ which we write as $12 = 5x$. Then divide both sides by 5 to get $x = \frac{12}{5}$ or $2\frac{2}{5}$.

PRACTICE 11

State if each pair of ratios forms a true proportion.

1) $\dfrac{3}{6} = \dfrac{7}{14}$

2) $\dfrac{5}{10} = \dfrac{8}{14}$

3) $\dfrac{7}{2} = \dfrac{18}{5}$

4) $\dfrac{9}{4} = \dfrac{18}{8}$

Solve the proportion

5) $\dfrac{3}{9} = \dfrac{5}{p}$

6) $\dfrac{x}{3} = \dfrac{4}{2}$

7) $\dfrac{6}{9} = \dfrac{r}{7}$

8) $\dfrac{2}{n} = \dfrac{10}{9}$

9) $\dfrac{5}{x} = \dfrac{2}{3}$

10) $\dfrac{8}{x} = \dfrac{4}{5}$

11) $\dfrac{b}{4} = \dfrac{6}{7}$

12) $\dfrac{4}{5} = \dfrac{v}{3}$

13) $\dfrac{9}{k} = \dfrac{2}{5}$

14) $\dfrac{n}{5} = \dfrac{2}{4}$

15) $\dfrac{2}{6} = \dfrac{5}{r}$

16) $\dfrac{4}{5} = \dfrac{n}{8}$

17) $\dfrac{9}{n} = \dfrac{2}{7}$

18) $\dfrac{3}{7} = \dfrac{9}{n}$

19) $\dfrac{r}{7} = \dfrac{2}{3}$

20) $\dfrac{5}{r} = \dfrac{6}{2}$

METRIC SYSTEM

For this course, you will occasionally have to convert among metric units (centimeters, meters, milliliters, liters, grams, kilograms) which is not as difficult as the Customary system (inches, feet, pints, gallons, ounces, pounds). The metric system is all based on 10's so converting from one unit to another just means moving the decimal point to the left or right.

In the metric system meters (m) are used for measuring length, liters (L) are used for measuring volume and grams (g) are used for measuring mass. We can put prefixes on the front to make the units bigger or smaller. For example, the weight of a paper clip is about one gram (g), but to measure your own weight you would probably use kilograms (kg). Soda is usually sold in two-liter (L) bottles. One milliliter (mL) mL of water weighs one gram. A meter stick which is 100 centimeters is roughly the same length as a yard stick. If you are going to participate in a 5K run, you should know the K stands for kilometers (km), so you're going to run roughly 3.1 miles. Here is a table of metric unit prefixes:

Prefix	Symbol	Multiplier	
exa-	E	10^{18}	1,000,000,000,000,000,000
peta-	P	10^{15}	1,000,000,000,000,000
tera-	T	10^{12}	1,000,000,000,000
giga-	G	10^{9}	1,000,000,000
mega-	M	10^{6}	1,000,000
kilo-	k	10^{3}	1,000
hecto-	h	10^{2}	100
deka-	da	10	10
deci-	d	10^{-1}	0.1
centi-	c	10^{-2}	0.01
milli-	m	10^{-3}	0.001
micro-	μ	10^{-6}	0.000001
nano-	n	10^{-9}	0.000000001
pico-	p	10^{-12}	0.000000000001
femto-	f	10^{-15}	0.000000000000001
atto-	a	10^{-18}	0.000000000000000001

Some of these prefixes you may have heard before, especially if you are buying a computer or reading some scientific papers. You can buy a terabyte hard drive which is bigger than a 100 gigabytes drive. You may have medicine measured in milligrams or micrograms, and you better hope that your nurse knows the difference. Nanotechnology is very, very small. In this course we will focus on milli-, centi-, base units, and kilo- and you will be able to convert between each of them.

To memorize the main metric conversions, we will use the mnemonic device:

KING HENRY DIED BY DRINKING CHOCOLATE MILK

KING	KILO- kilometers, kiloliters, kilograms
HENRY	HECTO-
DIED	DEKA-
BY	Base units: meters (m), liters (L), grams (g)
DRINKING	DECI-
CHOCOLATE	CENTI-
MILK	MILLI-

K   H   D   B   D   C   M

If we convert from one unit to another, we use the chart above to determine where we are starting, where do we end, then count the how many spaces we move to the left or right. Every space moved to the left multiplies by 10 while every space moved to the right divides by 10.

ARITHMETIC

Ex. 1 Convert 12 cm to km. Kilometers are bigger units than centimeters. When we go from a smaller unit to a bigger unit we need less of the bigger unit, the number should get smaller. The KHDBDCM shows us how many spaces to move the decimal. Since we start at the C and move left 5 spaces to the K, the decimal place should move 5 places to the left as well.

12 cm to km

K H D B D C M

0.00012 cm to km

12 cm = 0.00012 km

Ex. 2 Convert 200 L to mL. Milliliters are smaller than liters. When we go from a bigger unit to a smaller unit we need more of the smaller unit, the number should get bigger. Again, KHDBDCM shows us how many spaces to move the decimal. Since we start at the B (for base unit) and move right 3 spaces to the M, the decimal place should move 3 places to the right as well.

200 L to mL

K H D B D C M

200.000 L to mL

200 L = 200,000 mL

PRACTICE 12
Convert metric units

1. 20 mL to L
2. 4 g to cg
3. 40 mg to cg
4. .04 kg to g
5. .2 L to mL
6. .04 km to cm
7. 4.2 mg to g
8. 2400 mg to kg
9. 5.62 m to cm
10. 1.86 m to mm
11. 7,180 mm to m
12. 88.14 km to m
13. 54.41 L to mL
14. 48,710 mL to L
15. 8.55 g to mg
16. 69,500 mg to g

17. 7.88 kg to mg

18. 200 cm to km

19. 2 L 40 mL to mL

20. 3 L 200 mL

SOLVING PERCENT PROBLEMS

To solve percent word problems, we will translate the words into a mathematical equation to solve. Translate keywords into math language:

- "of" means multiply
- "is" means =
- "what" tells us to use a variable like x or n
- If you see "what percent" use p and remember to convert answer to percent
- If you see "what number" use n and remember your answer should just be a number
- Change percents into decimals

<u>Ex. 1</u> What percent of 82.5 is 20?

Translates into: $p \times 82.5 = 20$ or we can write as $82.5p = 20$. We solve by dividing both sides by 82.5 to get $p = \frac{20}{82.5}$ = .2424... which we change into a percent as 24.2%

<u>Ex. 2</u> 110% of 50.2 is what?

Translates into: $1.1 \times 50.2 = n$ Here we simply multiply 1.1 by 50.2 to get n, which comes out to 55.2

<u>Ex. 3</u> 26% of what is 3?

Translates into: $.26 \times x = 3$ or $.26x = 3$ We solve by dividing both sides by .26 to get $x = \frac{3}{.26} = 11.5$ which we leave as a decimal, because here we want a *number* not a percent.

PRACTICE 13

Solve each problem.

1) 70 is what percent of 115.1?

2) What percent of 115 is 141?

3) What is 53% of 140?

4) 174% of 12 is what?

5) 320% of what is 129?

6) 26 is 96% of what?

7) What percent of 122 is 12?

8) What is 78% of 5?

9) 270% of what is 111?

10) What percent of 123.7 is 121?

11) 38% of 15 hours is what?

12) 40 miles is what percent of 157 miles?

13) 82% of what is 68 m?

14) What is 95% of 132.6 miles?

15) 43.6 m is what percent of 156 m?

16) What is 32% of 13 miles?

17) What percent of $109 is $15?

18) 13% of what is 135.7 m?

19) 290% of 137 miles is what?

20) 54 tons is what percent of 14 tons?

PERCENT OF CHANGE

Percent of change is something we hear about all the time. Gas prices went up by 10%. Stock prices fell by 5%. College tuition has increased at an average rate of 3.1% over 10 years. Percent of change tells us how much something has changed relative to where it started from.

$$Percent\ of\ Change = \frac{Difference}{Original}$$

When giving percent of change, we must also include whether it was an *increase* or *decrease*.

ARITHMETIC

<u>Ex 1</u> Two weeks ago, the price of gas was $2.45. Now gas costs $2.65. What is the percent of change?

First, we find the difference, taking the higher number minus the lower number.

$2.65 - 2.45 = .20$ The difference is 20 cents.

Now divide the difference by the original price, which started at $2.45

$\frac{.20}{2.45} = .0816$ which we change to a percent 8.16% and we will round to the nearest tenth of a percent to get an 8.2% *increase*. We write increase because the price went up.

<u>Ex 2</u> During Christmas a popular video game console cost $450 now it is on sale for $300. What is the percent of change?

$450 - 300 = 150$ The difference is $150.

Now divide the difference by the original price, which started at $450

$\frac{150}{450} = .3333$ which we change to a percent 33.33% and we will round to the nearest tenth of a percent to get an 33.3% *decrease*. We write decrease because the price went down.

PRACTICE 14

Find each percent change. Round to nearest tenth of a percent. State if it is an increase or a decrease.

1) From 98 to 90

2) From 48 to 78

3) From 45 to 95

4) From 48 to 27

5) From 99 to 31

6) From 42 to 41

7) From 79 to 100

8) From 8 to 28

9) From $93 to $85

10) From 7 m to 29 m

11) From 33 hours to 11 hours

12) From 46 hours to 18 hours

13) From 73 tons to 1 ton

14) From 31 inches to 70 inches

15) From 32 hours to 20 hours

16) From $25 to $56

17) From 7 m to 19 m

18) From 78 ft to 31 ft

19) From 53 ft to 32 ft

20) From 15 hours to 9 hours

DISCOUNTS, MARKUPS, TAXES

We also use percents when we go shopping. We are familiar with sales, discounts, markups, and paying taxes on the things we buy and sell. I will go through two methods of solving these types of problems. Most of you will probably be familiar with the first method, but the second method is quicker and usually easier to do in your head.

Discounts

A discount or sale means that we will be paying less than the original price.

Ex. 1 A dress is originally $32 but is now on sale for 30% off. What is the sale price?

Method 1: Find the amount of the discount 30% of $32 by multiplying $.3(32) = 9.6$ Now this is the discount, not the final amount. We need to subtract 9.6 from 32. $32 - 9.6 = 22.4$ The sale price is $22.40

Method 2: This method works by determining what percent we actually have to pay. If it is 30% off that means we must pay 70% of the original price. Subtract the discount percent from 100% (100-30=70). So, what is 70% of 32 $.7(32) = 22.4$ The final price is $22.40. Notice we get the same result, but we have less math work to do. It's also easier to estimate the answer in our head when we are out shopping .7(30) =21 so $21 would be an estimate of the price.

Markups and Taxes

A markup or tax means that we will be paying more than the original price.

<u>Ex. 2</u> You go to buy a video game for $50.99 and sales tax is 7%. What is the sale price?

Method 1: Find the amount of the tax 7% of $50.99 by multiplying $.07(50.99) = 3.5693$ Now this is the tax, not the final amount. We need to add 3.5693 to 50.99. 50.99+3.5693= 54.5593 The sale price is rounded to the nearest cent (hundredths) to $54.56

Method 2: This method works by determining what percent we actually have to pay. If it is 5% more that means we pay 5% of the original price plus we have to pay ALL (100%) of the original price. Add the tax percent to 100% (100+5=105). So, what is 105% of 32 $1.05(50.99) = 53.5395$ The final price is $53.54. Again, we get the same answer.

Combining Discount with Markup/Tax

<u>Ex. 3</u> A pair of shoes is originally $100 but is marked up by 50% then on sale for 30% off. What is the final price?

Method 1: Find the amount of the markup 50% of $100 by multiplying $.5(100) = 50$ Add this to the original $50 + 100 = 150$. Then find the discount of 30% of 150 by $.3(150) = 45$. We need to subtract 45 from 150. $150 - 45 = 105$ The sale price is $105

Method 2: If it is marked up by 50% off that means we must pay 150% of the original price. We also have a discount of 30% off, so that means we pay 70% of the price. We can do both steps at the same time. $.7(1.5)(100) = 105$ The final price is $105.

Notice that whichever method you choose, the order that you apply the discount and markup does not matter. Discount first, then markup versus markup first, then discount; either way it ends up the same.

What is the total percent off if you have a 50% discount and then a 30% discount? If you thought 80% then you have fallen for a trap. Many retail stores will have multiple sales on an item, making it appear that you will save much more than you really are. Let's do the math. Suppose the item is $100. If it is 50% off, now the item is $50. The 30% off is 30% off that price not the original. You pay 70% of $50 which is $35. The total savings is $65 out of $100. That's a 65% discount, not 80% which you may have thought originally.

PRACTICE 15

Find the final price of each item.

1. Original price of bike: $119.99   Discount: 40%

2. Original price of camera: $59.90   Discount: 25%

3. Original price of lamp: $34.90   Discount: 30%

4. Original price of swimsuit: $44.99   Discount: 20%

5. Original price of shoes: $114.99   Markup: 45%

6. Original price of shirt: $15.95   Markup: 33%

7. Original price of shorts: $24.90   Markup: 40%

8. Original price of medication: $50.95   Markup: 50%

9. Original price of video game: $59.99   Tax: 4%

10. Original price of necklace: $24.95   Tax: 7%

11. Original price of phone: $245.95   Tax: 6%

12. Original price of bedding: $65.95   Tax: 4.5%

13. Original price of robot: $99.99 Discount: 20% Tax: 5%

14. Original price of laptop: $699.95 Discount: 15% Tax: 6%

15. Original price of fryer: $140.90 Discount: 40% Tax: 4%

16. Original price of watch: $124.95 Discount: 30% Tax: 4.5%

SIMPLE INTEREST

Interest is the fee that you pay for borrowing money or money that is earned when you lend money. Simple interest is the simplest way to calculate interest and can provide a reasonable estimate for many situations. However, most loans and credit cards companies calculate interest by compound interest. In compound interest, the interest is added to the principal (original amount) and then earning interest on the interest. Our focus for this course will be solving problems with simple interest, using the formulas:

$A = Prt$ which states the amount of interest = principal x annual interest rate x time in years

$M.V. = P + A$ maturity value = principal + interest

Maturity value is the total to be paid back at the end, principal and interest together.

$$monthly\ payment = \frac{Maturity\ Value}{number\ of\ months}$$

Ex. 1 You decide to invest $2,000 in a certificate of deposit (CD), with 4% interest for 3 years. What will be the maturity value?

First, we find the interest $A = prt = (2000)(.04)(3) = 240$ Then we add the interest to the principal $2000 + 240 = 2240$. The maturity value will be $2,240.

Ex. 2 Find the monthly payment for a loan of $3,500 with an interest rate of 7.5% for 5 years.

First, we find the interest $A = prt = (3500)(.075)(5) = 1312.5$ Then we add the interest to the principal $3500 + 1312.5 = 4812.5$. The maturity value will be $4812.50. Five years is 5(12) = 60 months. The monthly payment is $\frac{4812.5}{60} = 80.2083$. We round to the nearest cent $80.21

PRACTICE 16

Use simple interest to maturity value and monthly payment

1. $12,500 at 5% for 3 years

2. $14,600 at 6% for 7 years

3. $2,400 at 7% for 3¾ years

4. $1,700 at 6% for 2½ years

5. $348,600 at 4.2% for 30 years

6. $16,700 at 5.8% for 5 years

7. $2,300 at 4.5% for 18 months

8. $14,600 at 6.5% for 30 months

9. $1,500 at 7.25% for 21 months

10. $24,000 at 5.35% for 60 months

SOLVING PROPORTION WORD PROBLEMS

When setting up proportions from word problems we have to be careful that both sides are set up the same way. For example if we compare apples to oranges on one side, then on the other side we need to compare apples to oranges $\frac{apples}{oranges} = \frac{apples}{oranges}$

We would not want apples to oranges and then oranges to apples $\frac{apples}{oranges} \neq \frac{oranges}{apples}$

This should become clearer as we work through examples.

Ex. 1 The currency in Kuwait is the Dinar. If 6 dinars is equal to $20, then how many dinars is $14.

The first piece of information we can use on one side of the proportion. Once one side is set up, that determines how we set up the second side. $\frac{\$20}{6\ dinars} = \frac{\$14}{x}$ On the left side we put dollars on top and dinars on the bottom, so on the right side, dollars have to go on top. Including the labels helps to put the numbers in the correct space. "Cross-multiply and divide" to solve.

$\frac{20}{6} = \frac{14}{x}$ becomes $20x = 6(14)$ then $20x = 84$ so $x = 4.2$ Therefore, $14 is equal to 4.2 dinars.

Ex. 2 A painting is 8 inches wide and 12 inches tall. If the painting is enlarged to be 20 inches tall then how tall will it be?

For geometry problems such as this drawing a picture can be helpful.

$\frac{8\ wide}{12\ tall} = \frac{x}{20\ tall}$ On the left side we put the width on top and height on the bottom, so on the right side, we do the same.

$\frac{8\ left\ width}{x\ right\ width} = \frac{12\ left\ height}{20\ right\ height}$ We could set it up this way too. On the left side we put the left width on top and right width on the bottom, so on the right side, we put the left height on top and right height on the bottom. There can be more than one correct way to set up a proportion. Just make sure you are comparing the same way on both sides. Check that answers seem reasonable.

$\frac{8}{12} = \frac{x}{20}$ becomes $12x = 160$ so $x = \frac{160}{12} = 13\frac{1}{3}$ or $13.\overline{3}$ inches. This is a reasonable answer since we expect it to be bigger than the original 8 inches, but smaller than the new height of 20 inches.

Ex. 3 The pair of figures is **similar**, the sides are in proportion to each other.

 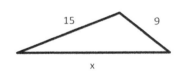

Even though these triangles are similar they are not oriented in the same way. We also have more information than we need. To set up a proportion we need 4 pieces of information (one of which should be our missing side x). It will be helpful to mark the picture. First find the x which is the longest side on the triangle to the right, I mark that with a single slash. We also mark the longest side on the left triangle with a single slash. Then we need to mark two other sides. We mark the smallest sides with two slashes on each triangle. Now we are ready to set up our proportion. One way to set it up

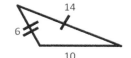 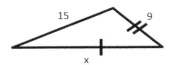

is $\frac{14}{6} = \frac{x}{9}$. Notice here we have the large side 14 to small side of 6 for the triangle on the left compared to the large side x to small side of 9 for the triangle on the right $6x = 126$. Then $\frac{6x}{6} = \frac{126}{6}$ which gives $x = 21$. 21 seems like a reasonable answer since by the picture we expect it to be the largest side and larger than 15[24].

PRACTICE 17

1. 10 avocados cost $12.70 How much are 4 avocados?

2. One package of blueberries cost $2.50. How many packages can you buy with $10?

3. Matthew bought 12 candy bars for $18. How many can he buy with $9?

4. John bought 14 oranges for $6.50. How much is 6 oranges?

5. The exchange rate is 1 euro is equal to $1.12. How many euros is $15.68?

---

[24] The side x also has to be smaller than 24 by the triangle inequality which says that any side of a triangle is smaller than the sum of the other two sides 15 + 9 = 24.

6. 20 British pounds is equal to $26.12. How many dollars is 12 pounds?

7. 300 Indian Rupees is $4.32. How many dollars is 160 Indian Rupees?

8. $1 USD is equal to $1.34 Canadian dollars. How many US dollars is something that costs 7.50 Canadian Dollars?

9. Find the missing side

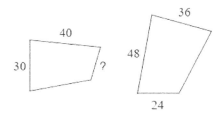

10. Find the missing side

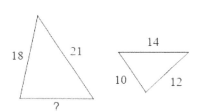

11. A picture is 5 inches in width by 8 inches long. It is enlarged so that height is 20 inches. What is the new width?

12. The Washington monument is 169 meters tall and 17 meters wide. A scale model is made to have a height of 30 centimeters. What is the width in cm?

GEOMETRY

We are now entering the second main section, geometry the branch that deals with the properties of points, lines, shapes, and solids. We will cover Euclidean[25], "flat", geometry as you may have learned in high school. Here's a table to summarize and visualize some of the basic geometric terms and definitions.

Euclid of Alexandria

Term	Definition	Picture
Point	A location in space, having 0 dimensions, no length, no width, no height. Represented by a dot and labeled by a capital letter	• A
Line	Has 1 dimension, length but no width or height. A line extends forever in two directions. Represented by a line with arrows at both ends. Labeled by a script letter or by two points on the line.	ℓ $\overleftrightarrow{AB}$
Line segment	Part of a line with two endpoints, a beginning and an end.	$\overline{AB}$
Ray	Part of a line with one endpoint and goes on forever in the opposite direction.	$\overrightarrow{AB}$
Plane	Flat 2-dimensional object with length and width like an infinite chalkboard.	P
Polygon	A flat 2-dimensional shape formed by lines connected together. Common polygons include triangles, stars, rectangles, hexagons, etc.	

---

[25] Euclid codified the knowledge of ancient geometry in his book the *Elements* written around 300 BC (Norton, 2017)

Solid	A 3-dimensional object having length, width, and height. Common solids include cubes, cones, cylinders, spheres, etc.	
Parallel lines	Lines that never cross. They are always the same distance apart. They also have the same **slope** (discussed later). Parallel lines are marked by a small triangle placed on the line.	$\ell \parallel m$
Intersecting lines	In Euclidean plane geometry, lines are either parallel or they intersect in one point.	

TYPES OF ANGLES

An angle is formed from two rays with a common endpoint called the vertex. We can

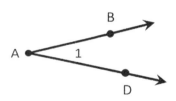

label an angle by the points on the rays and the vertex. For example, here's an angle and we can name it several ways:

$\angle BAD, \angle DAB, \angle A, \angle 1$. When we measure an angle, we measure how much of a turn we would have to make from ray to the second ray. A complete turn is 360°, half a turn is 180°, and a quarter turn is 90°.

Knowing what some common angles measures look like will help in estimating your answers.

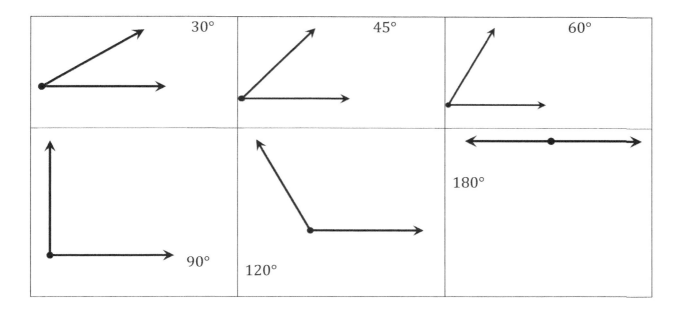

There are several types of angles and angle relationships that we will work with.

TERM	DEFINITION	PICTURE EXAMPLE
Right angle	90° angle marked by a square at the vertex	
Perpendicular lines	Lines that form a right angle. Symbol for perpendicular is ⊥	$\ell \perp m$
Acute angle	less than 90 degrees	
Obtuse angle	more than 90 degrees	
Straight angle	Equal to 180 degrees. A straight angle forms a straight line.	

GEOMETRY

Complementary angles	Two angles that add up to 90 degrees	
Supplementary angles Linear Pair	Two angles that add up to 180 degrees	
Vertical angles	Two angles formed by intersecting lines. Vertical angles are congruent (same)	
Adjacent angles	Angles that are "next" to each other. Adjacent angles have the same endpoint and have a common side.	

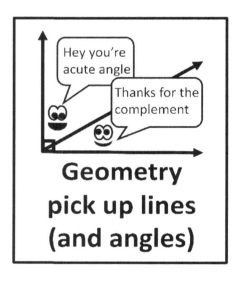

So far, we have four different angle relationships: complementary, supplementary, vertical and adjacent. For complementary look for the little square to let you know there is a right angle. For supplementary angles, look for two angles that make up a straight line. Vertical angles are formed from two intersecting lines. Adjacent angles are just next to each other and we need extra information to find a missing angle.

Ex. 1 $m\angle HIK = 70°$ and $m\angle KIJ = 86°$. Find $m\angle HIJ$.

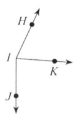

In this example we are given the measure of angles HIK and KIJ which are both the smaller angles. We need to find angle HIJ which is the large angle. We are given two parts and need to find the whole. So, we will add the parts together to get the whole HIJ. 70 + 86 = 156.

Page | 68

Ex. 2 $m\angle STG = 40°$ and $m\angle STU = 132°$. Find $m\angle GTU$.

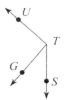

In this second example we are given the measure of angle STG which is a small angle and STU which is the larger angle. We need to find angle GTU which is another small angle. We are given one part and the whole. We need to find the other part. So, we will subtract the part from the whole to get GTU. 132 – 40 = 92.

Ex. 3 Classify each angle based on appearance

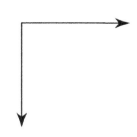

Even though this angle is not marked, by appearance it looks like 90 degrees. We can't say it looks acute or obtuse, so just by appearance the best answer is that it is a right angle.

Ex. 4 Name the relationship and find the missing angle.

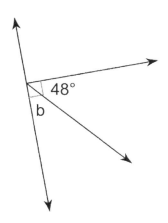

These two angles share a common side and a common vertex so they are adjacent, but we are looking for the best description of the two angles. Notice the small square in the corner tells us we have a right angle that is formed by two smaller angles. The angles must be complementary. Therefore, we can find the missing angle by subtracting 48 from 90 to get 42 degrees.

Ex. 5 Name the relationship and find the missing angle.

Focus on the lines that make up the given angle of 56 degrees and angle b. We can see that these lines intersect each other to form the angles. The two angles are across from each other which make them vertical angles. Vertical angles are always congruent so b must be equal to 56 degrees.

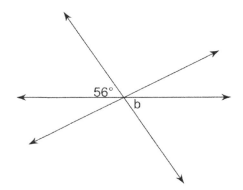

Ex. 6 Name the relationship and find the missing angle.

This example looks similar to the previous with two intersecting lines, however, the 53 degree angle and b are side by side, so you may think they are adjacent which is true, but there is a better description for these two angles. The two angles form a straight 180 degree angle, therefore they are supplementary. Subtract 53 from 180 to get angle b is 127 degrees.

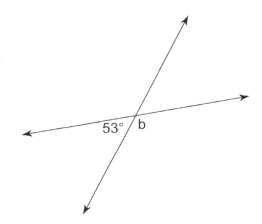

Ex. 7 Name the relationship and find the missing angle.

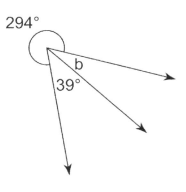

This example has two angles side by side. They are not supplementary or complementary, so they are just adjacent. Notice that we have two angles given to us, 39 and 294. Having the exterior angle marked like 294 is here is a clue that we have adjacent angles. To find the missing angle we must subtract both 39 and 294 from 360 degrees to find what's left. We get that the angle b is 27 degrees.

PRACTICE 18

1) Find $m\angle LMG$ if $m\angle LMN = 134°$ and $m\angle GMN = 100°$.

2) $m\angle LBC = 106°$ and $m\angle ABC = 176°$. Find $m\angle ABL$.

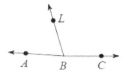

3) $m\angle RGF = 39°$ and $m\angle HGR = 116°$. Find $m\angle HGF$.

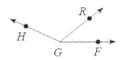

4) Find $m\angle LMN$ if $m\angle HMN = 50°$ and $m\angle LMH = 98°$.

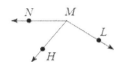

Classify each angle as acute, obtuse, right, or straight.

5)

6)

7)

8)

9) 144°

10) 90°

11) 180°

12) 25°

GEOMETRY

Name the relationship and find the missing angle

13)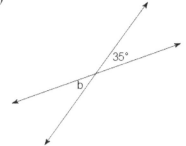

14)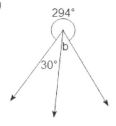

15)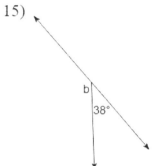

16)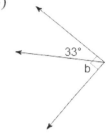

17)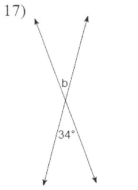

18)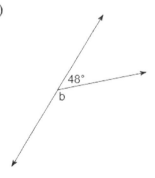

19)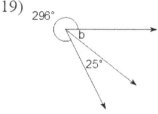

20)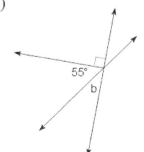

PARALLEL LINES AND ANGLE RELATIONSHIPS

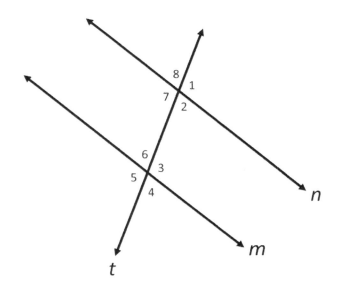

When two parallel lines are crossed by a third line (transversal), we get more angles and angle relationships. In the diagram to the right, line *m* is parallel to line *n*; we can write this as *m* ∥ *n*. Line *t* is the transversal because it cuts across the two other parallel lines. How many angles are formed? There are eight new angles formed and four special relationships which we discuss now:

Alternate Interior Angles (Alt. Int. ∠) – interior means inside the parallel lines and the alternate refers to being across the transversal. In this picture, ∠7 and ∠3 are Alt. Int. ∠6 and ∠2 are Alt. Int. It appears that alternate interior angles are the same measure, and they are. Alternate interior angles are **congruent** (same, ≅). We can say ∠7 ≅ ∠3.

Alternate Exterior Angles (Alt. Ext. ∠) – exterior means outside and the alternate means across the transversal. In this picture, ∠8 and ∠4 are Alt. Ext. ∠1 and ∠5 are Alt. Ext. Alternate exterior angles are also congruent.

Corresponding Angles (Corr. ∠) – corresponding means matches up to another angle. Imagine taking ∠8 at the top and if you were to slide it up, down, left, right (without flipping or rotating) which angle would it match up to? ∠8 and ∠6 are corresponding. The other corresponding pairs are ∠1 and ∠3, ∠2 and ∠4, ∠7 and ∠5. Corresponding angles as you may have guessed are also congruent.

Same-Side Interior Angles (S-S int. ∠) also sometimes called **consecutive interior** – so these are pairs of angles that are both inside the parallel lines and on the same side of the transversal. ∠7 and ∠6 are S-S Int. ∠3 and ∠2 are S-S Int. Are the angles in these pairs congruent? Nope, but they do have a special property: same-side interior angles are supplementary. So, $m∠7+m∠6 = 180°$ and $m∠3+m∠2 = 180°$.

Seems like a lot to memorize, but if you do happen to forget the names, remember this: **Big = Big, Small = Small, Big + Small =180.** That is, if you take any acute angle, it will be congruent to any other acute angle when you have 2 parallel lines and transversal. Any big, obtuse angle will be equal to any other big angle. If you take any big angle and add any small angle, they will add up to 180 degrees.

<u>Ex. 1</u> Identify the relationship and find the missing angle.

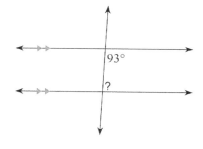

In the diagram, the two angles are inside the parallel lines making them interior angles. They are also on the same side of the transversal. This makes them same-side interior angles which should add up to 180 degrees. 180 – 93 = 87.

<u>Ex. 2</u> Identify the relationship and find the missing angle.

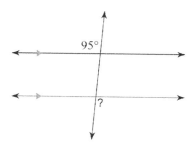

Here, the two angles are outside the parallel lines making them exterior angles. They are also on the opposite sides of the transversal. This makes them alternate exterior angles which are congruent. This missing angle must be 95 degrees.

GEOMETRY

Ex. 3 Identify the relationship and find the missing angle.

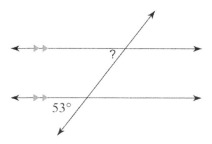

In this final example, one angle is inside the parallel lines while the other angle is outside the parallel lines. This means they cannot be interior or exterior angles. The only option left is corresponding angles. See how the two angles are open in the same direction and match up? Corresponding angles which are congruent. This missing angle must be 53 degrees.

> PARALLEL LINES HAVE SO MUCH IN COMMON, IT'S A SHAME THEY WILL NEVER MEET.

PRACTICE 19

Name the relationship and find the measure of each angle indicated.

1)
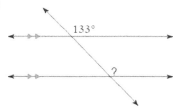

2)
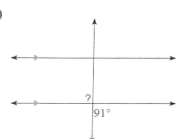

3)
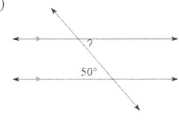

4)
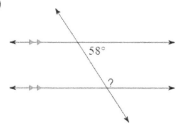

5)
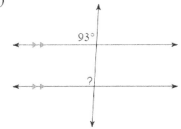

6)
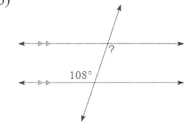

7)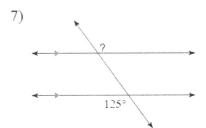

8)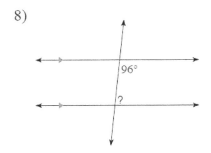

9)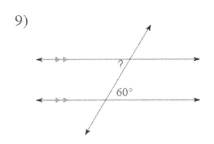

10)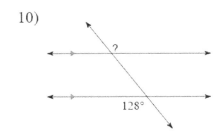

TRIANGLES

A triangle is a shape with 3 straight sides. Triangles are very important in geometry and the real world, building bridges and houses, flying planes, using GPS to navigate, measuring the heights of large trees, finding distances to far away objects, etc. For a triangle, all the angles inside add up to 180°.

In the picture to the right, the angles must add up to 180 degrees, so the missing angle must be $x = 180 - 50 - 70 = 60$

We can classify triangles by their angles or their sides. Here's two tables to show how:

	CLASSIFY TRIANGLES BY ANGLES		
TERM	Right	Acute	Obtuse
DEFINITION	Triangle with 1 right angle	Triangle with all acute angles	Triangle with 1 obtuse angle
PICTURE			

GEOMETRY

	CLASSIFY TRIANGLES BY SIDES		
TERM	Equilateral	Isosceles	Scalene
DEFINITION	Triangle with 3 equal sides (it will also have 3 equal angles, equiangular)	Triangle with 2 equal sides (base angles will also be equal)	Triangle with no equal sides
PICTURE	We put marks on sides to show ≅		

We can classify any triangle by both its angle and side

Ex. 1 Classify the triangle. First, we look at the angles. Since it has a square at one vertex that tells us it has a right angle there. Therefore, it is a right triangle. Two sides are marked with a slash, so those two sides must be congruent, making it isosceles. This is a RIGHT ISOSCELES.

Ex. 2 Classify the triangle. First, we look at the angles. Since it has all angles smaller than 90 degrees, it must be acute. There are no sides marked the same. You might think the left and right sides look the same, but geometry is all about proving something to be true. These sides are not marked, so we cannot assume they are congruent. This is an ACUTE SCALENE.

Ex. 3 Classify the triangle. First, we look at the angles. Since it has all angles smaller than 90 degrees, it must be acute. All the sides are marked the same, making it equilateral. ACUTE EQUILATERAL. In fact,

since all the angles are also equal, and the angles add up to 180 degrees, each angle will always be 60 degrees. All equilateral triangles will be acute.

Ex. 4 Classify the triangle. First, we look at the angles. It has an obtuse angle. Two sides are marked the same, making it isosceles. OBTUSE ISOSCELES.

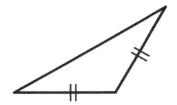

PRACTICE 20

Sketch an example of the type of triangle described. Mark the triangle to indicate what information is known. If no triangle can be drawn, write "not possible."

1) acute isosceles

2) scalene isosceles

3) right scalene

4) equilateral

5) right equilateral

6) obtuse scalene

Classify each triangle by its angles and sides. Equal sides and equal angles, if any, are indicated in each diagram.

7)

8)

9)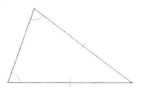

10)

11)

12)

13)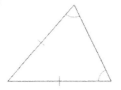

14)

15)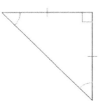

16)

17)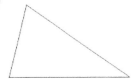

18)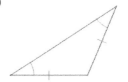

FINDING MISSING ANGLES IN A TRIANGLE

All the angles in a triangle add up to 180 degrees

$\angle A + \angle B + \angle C = 180$

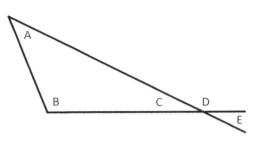

When a side is extended, we have supplementary angles $\angle C + \angle D = 180$

An exterior angle is equal to the sum of the two opposite interior angles

$\angle D = \angle A + \angle B$

Sometimes vertical angles will be used if you have two lines crossing to form an X. Vertical angles are equal $\angle C = \angle E$

Ex. 1 Find the missing angle

We have some extended lines, so we will have to use some supplementary angles and triangle angle sum. When working these problems try to fill in all the angles we can step by step. We need to know all the angles inside the triangle. First, we start with the 147. Its supplement is 180-147=33. Now we have two angles inside a triangle. All 3 angles must add up to 180 degrees. 180-118-33=29. Once we have that our mystery angle is supplementary to 29 so it must be 180-29=151. Of course, we could have skipped a couple steps and used the exterior angle formula, 33+118=151.

GEOMETRY

Ex. 2 Find the missing angle.

This is example is more complicated and involves two triangles, but we apply the same properties and formulas. We want to fill in all the missing angles of the triangles. We can start on the right side where the little square shows we have a right angle. We can use vertical angles to get 90 degrees in the triangle to the right. There is an angle at the top left which is supplementary to 150 which is 30 degrees. We subtract 180 – 30 – 65 = 85, the remaining angle in the left triangle is 85 degrees. We have 3 angles which will add up to 180 degrees. 180 – 85 – 55 = 40. Now that we have two angles in the right triangle, we can subtract those from 180. 180 – 40 – 90 = 50. So, after all that work, we have the answer is 50 degrees. Which seems like a reasonable answer since it appears to be acute angle smaller than most of the other labeled angles.

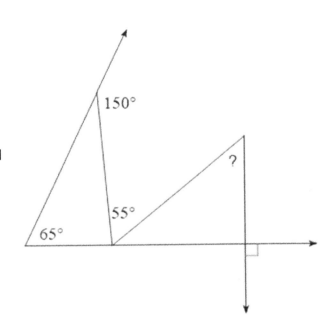

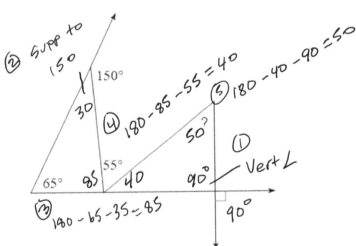

Page | 81

PRACTICE 21

1)

2)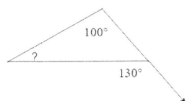

3)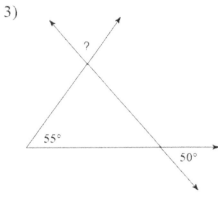

4)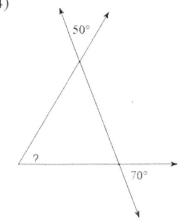

5)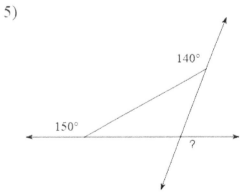

6)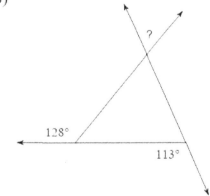

7)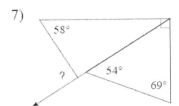

8)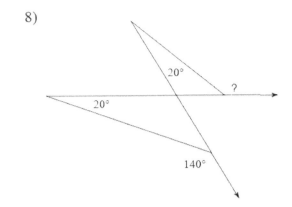

9)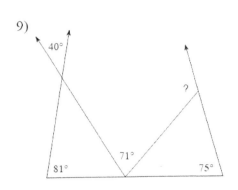

10)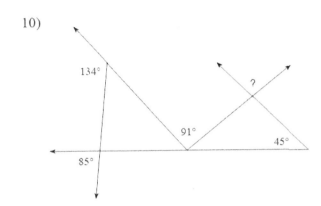

11)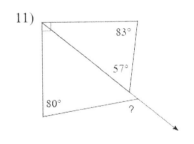

12)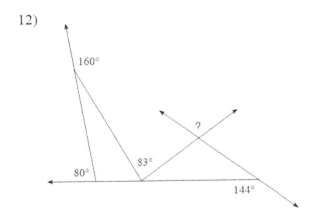

CLASSIFYING QUADRILATERALS

Quadrilaterals are shapes that have 4 sides. There are some special quadrilaterals for you to classify:

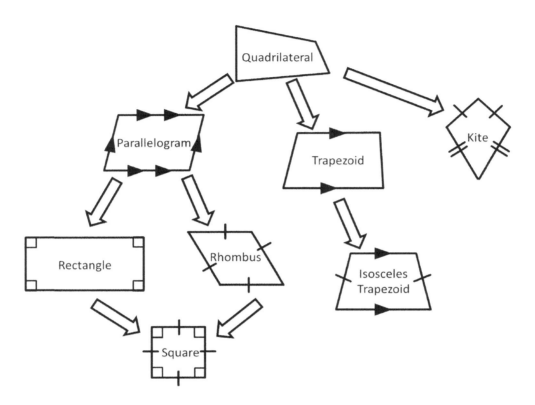

Parallelogram – two pairs of parallel lines

Rectangle – a parallelogram with 4 right angles

Rhombus – a parallelogram with 4 equal sides

Square – both a rectangle and rhombus, all sides equal, all angles equal

Trapezoid – only one pair of parallel sides

Isosceles Trapezoid – a trapezoid with 2 congruent sides

Kite – has two pair of consecutive sides that are congruent

GEOMETRY

Ex. 1 State the most specific name for each figure

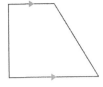
In this diagram we have the top and bottom lines parallel since they have the arrow marking on them. Since there is only one pair of parallel lines, we have ourselves a trapezoid.

Ex. 2 State the most specific name for each figure

Here the slashes tell us that all the sides are congruent. Four equal sides give us a rhombus, which of course is a parallelogram (and a quadrilateral) but we are to only give the most specific name – rhombus.

Ex. 3 State the most specific name for each figure

In this example, we have a pair of parallel lines, so it is a trapezoid, but notice it also has slashes on the top and bottom marking that those two sides are congruent. Therefore, we have an isosceles trapezoid.

Ex. 4 State the most specific name for each figure

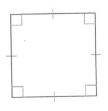
This quadrilateral has four equal sides and four equal right angles so it must be a square.

GEOMETRY

Ex. 5 State the most specific name for each figure

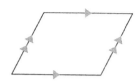

This figure has the top and bottom sides marked parallel and it has the left and right sides also marked parallel by the arrows. Two pairs of parallel lines give us a parallelogram.

Ex. 6 State the most specific name for each figure

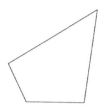

This four-sided figure has no markings at all. It has no special properties besides having four sides, so it must be a quadrilateral.

Ex. 7 State the most specific name for each figure

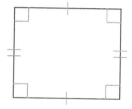

If you're not careful and just basing off appearance you may think this is a square, but the markings tell us we have four equal right angles, but the sides are not all equal. The top and bottom have one slash while the left and right have two slashes. Four equal right angles means this has got to be a rectangle.

Ex. 8 State the most specific name for each figure

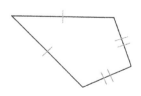

This last example has two pairs of congruent sides, but they are not opposite each other. The equal sides are adjacent, side by side, so this is the kite. You're ready to fly!

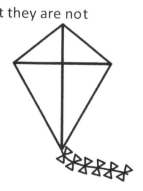

PRACTICE 22

State the most specific name for each figure.

1)

2)

3)

4)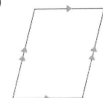

5)

6)

7)

8)

9)

10)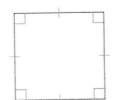

CLASSIFYING POLGONS

Number of Sides	Name	Picture
3	Triangle	△
4	Quadrilateral	□
5	Pentagon	⬠
6	Hexagon	⬡
7	Heptagon	
8	Octagon	⯄
9	Nonagon	
10	Decagon	
12	Dodecagon	

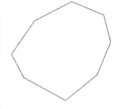

Ex. 1 Name the polygon

Naming polygons is easy just count the sides and memorize the table above. This polygon has eight sides which makes it an octagon.

PRACTICE 23
Name the polygon

1)

2)

3)

4)

5)

6)

7)

8)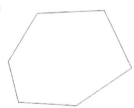

9)

10)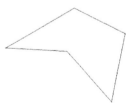

11)

12)

CLASSIFYING SOLIDS

Every physical object we encounter in everyday life is a solid. Solids are 3-dimensional objects that have width, depth, and height. There are two main types and then a few special cases. The two main types we will deal with here are prisms and pyramids. Both are made from polygons (straight edges, flat sides).

A **prism** is created by having two exact copies of a polygon, called the **bases**. Then create a box by connected the bases with rectangles. The bases don't necessarily have to be on the top and bottom, they could be on the sides or front and back. Find them first to name your prism correctly.

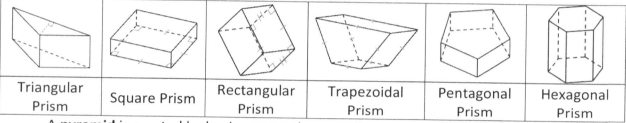

A **pyramid** is created by having one polygon, called the **base**. Then create triangular sides that meet at a point called the apex. Again, to name a pyramid, first find the base.

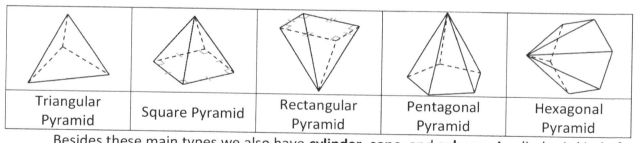

Besides these main types we also have **cylinder**, **cone**, and **sphere**. A cylinder is kind of like a prism (box) but instead of polygon bases it has two circular bases. A cone is like a pyramid with a circular base. And a sphere is the mathematical name for a ball.

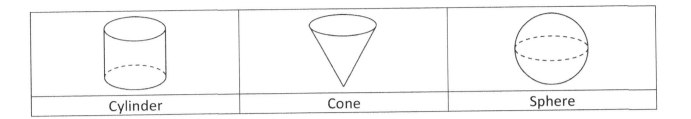

| Cylinder | Cone | Sphere |

PRACTICE 24
Identify the solid

1)

2)

3)

4)

5)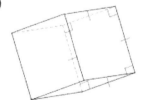

6)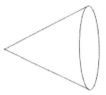

7)

8)

9)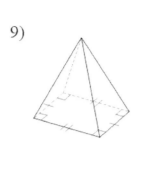

10)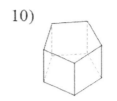

11)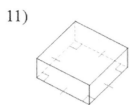

12)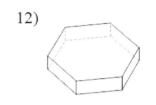

13)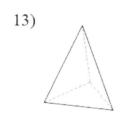

14)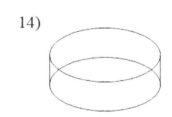

15)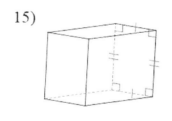

16)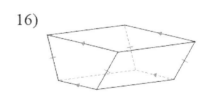

17)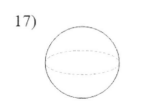

18)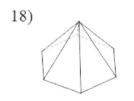

19)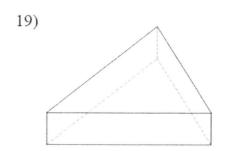

20)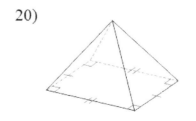

GEOMETRY

INTRODUCING COORDINATE GEOMETRY

The coordinate plane or Cartesian Plane (named after French mathematician René Descartes) gives us a way of quantifying space and relating geometry to algebra. A coordinate plane combines two numbers lines. The x-axis goes left and right and the y-axis goes up and down. The axes split a plane into 4 parts called quadrants which are labeled with roman numeral starting in the top right and going counterclockwise. Where the two axes intersect is the point (0,0) or the origin. Points are given coordinates (x, y). The x coordinate tells us how far to the right (positive) or left (negative) to go, while the y coordinate tells us how far up (positive) or down (negative) to go.

René Descartes
1596-1650

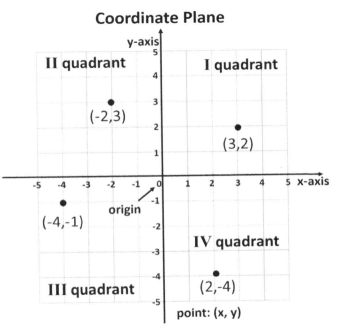

Ex. 1 State the coordinates of each point

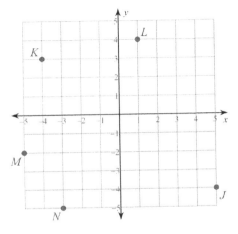

Remember the x coordinate goes first then the y coordinate. For point J we have to go right 5 spaces and down 4 so J is at (5, -4). For point K we have to go left 4 spaces and up 3 so K is at (-4, 3). L is at (1, 4). M is at (-5, -2). N is at (-3, -5).

Page | 93

PRACTICE 25

Plot each point.

1) $D(-9, 0)$ $E(-7, 4)$ $F(2, -1)$
   $G(8, -5)$ $H(-8, -3)$ $I(-5, -2)$
   $J(0, 6)$ $K(3, -4)$ $L(8, -10)$
   $M(5, 7)$

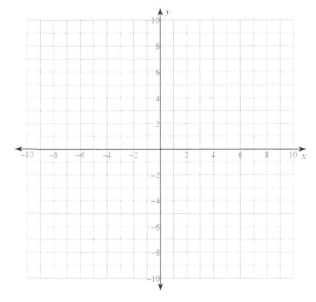

State the coordinates of each point.

2)

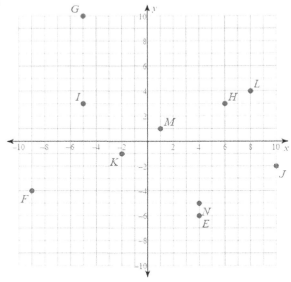

Blank Coordinate Plane for notes/examples

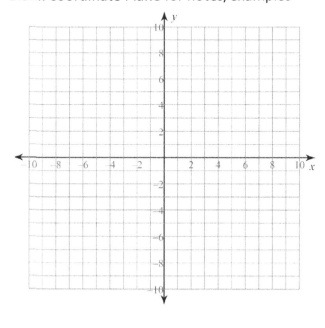

SNAKES ON A PLANE

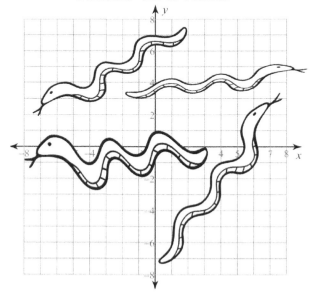

FINDING THE MIDPOINT

Suppose we have a line segment and we want to find the point that is halfway along it. We could measure the length and divide it by two, but what if we can't measure it directly and we only know the location of the endpoints? The midpoint formula gives us a way to find it. First, we will find the midpoint.

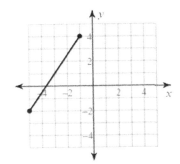

Ex. 1 Find the midpoint.

The first thing we will do is add a vertical line segment and horizontal line segment make a right triangle, see figure below. Count the horizontal distance which is 4 and the vertical distance is 6. Half of 4 is 2, while half of 6 is 3. Draw a vertical line halfway across and a horizontal line halfway down. We have the midpoint at (-3, 1). Using a picture helps a lot because we can just split up the distance across and distance down. This is how the midpoint formula works, by taking the average of the x's and the average of the y's.

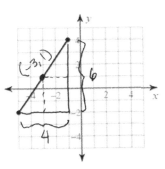

Suppose we have two points (x_1, y_1) and (x_2, y_2) then

$$Midpoint\ M = \left(\frac{x_1+x_2}{2}, \frac{y_1+y_2}{2}\right)$$

Here's the same problem worked using the formula. The endpoints are $(-5,-2)$ and $(-1,4)$.

Below the points, I label them $x_1, y_1, x_2,$ and y_2. Then substitute into the formula, and finish.

$(-5,-2)$ and $(-1,4)$
$x_1y_1x_2y_2$

$M = (\frac{x_1+x_2}{2}, \frac{y_1+y_2}{2})$
$= (\frac{-5+-1}{2}, \frac{-2+4}{2})$
$= (\frac{-6}{2}, \frac{2}{2}) = (-3,1)$

PRACTICE 26
Find the midpoint.

1)
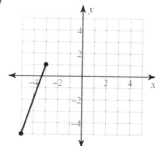

2)
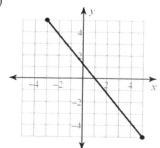

3)
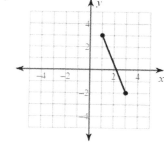

4)
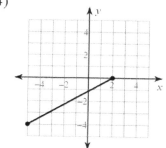

5) $(-14, 14), (8, 0)$

6) $(-4, -4), (10, 14)$

7) $(-11, 7), (-11, -6)$

8) $(-5, -9), (-4, -13)$

9) $(4.5, -4.4), (7.2, 7)$

10) $(8.8, 2.6), (-8.3, 5.2)$

11) $\left(2, -\dfrac{2}{9}\right), \left(\dfrac{7}{4}, 4\right)$

12) $\left(\dfrac{12}{7}, 3\dfrac{1}{3}\right), \left(4\dfrac{4}{7}, 1\dfrac{1}{9}\right)$

PYTHAGOREAN THEOREM AND FINDING SIDES OF A RIGHT TRIANGLE

The Pythagorean theorem is one of the most widely known theorems in the history of mathematics with many geometric and algebraic proofs. There is evidence that it was known long before Pythagoras, but it is said that he gave the first formula proof of the theorem and it is attributed to him. The theorem is used to describe the relationship between the sides of a right triangle. If we take the sum of the squares of the legs, we get the square of the hypotenuse, longest side. [26]

PYTHAGORAS

Pythagorean Thm:
$a^2 + b^2 = c^2$

[26] (Wikipedia Contributors, 2019f)

Ex. 1 Find the missing side below

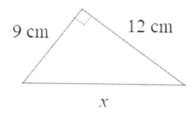

First identify the hypotenuse which is always opposite the right angle. The Square points to the hypotenuse x. Label the legs a and b, which are interchangeable. Then substitute into the Pythagorean theorem: $a^2 + b^2 = c^2$ to get $9^2 + 12^2 = x^2$. Simplify the right side, $81 + 144 = x^2$. Then $225 = x^2$. Finally take the square root of both sides to get $x = 15$. This answer makes sense since the hypotenuse should be the longest side.

Ex. 2 Find the missing side below

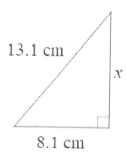

Here the hypotenuse is 13.1 and the legs are 8.1 and x. Use $a^2 + b^2 = c^2$ to get $x^2 + 8.1^1 = 13.1^2$. Simplify the right side, $x^2 + 65.61 = 171.61$. Here we subtract 65.61 from both sides to get $x^2 = 106$. Taking the square root to get $x = 10.2956$ which we can round to the tenths 10.3

PRACTICE 27

Find the missing side of each triangle. Round your answers to the nearest tenth if necessary.

1)
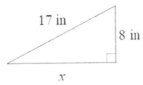

2)
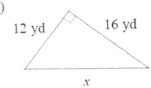

3)
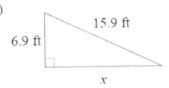

4)
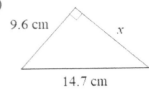

5)
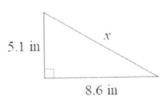

6)
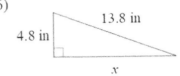

7)
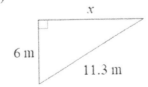

8)
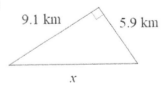

9)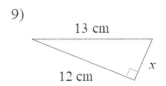

10)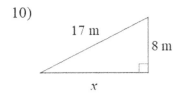

11)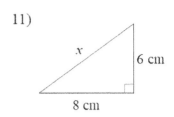

12)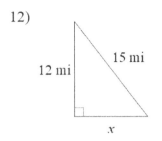

13)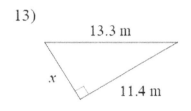

14)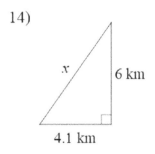

15)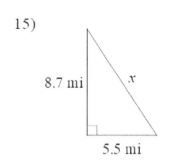

16)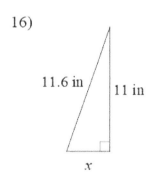

GEOMETRY

FINDING THE DISTANCE BETWEEN TWO POINTS

Suppose we want to find the distance between these two points or find the length of this line segment. If we have a graph, then we can do what we did with the midpoint problem and draw a triangle. How will this help? We create a right triangle and we can use the Pythagorean theorem. The distance we want to find is the length of the hypotenuse. The legs are 5 and 7. $5^2 + 7^2 = d^2$ and $25 + 49 = d^2$ Then $74 = d^2$. Taking the square root, we get $d = \sqrt{74} = 8.6023$ which rounds to 8.6

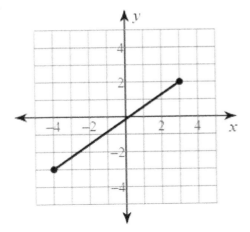

There is a distance formula, but it's pretty difficult to memorize and essentially, it's the same as the Pythagorean Theorem.

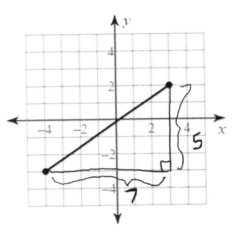

$$Distance = \sqrt{(x_1 - x_2)^2 + (y_1 - y_2)^2}$$

We will do the same problem, this time using the formula. We need the coordinates of the two points which are $(-4, -3)$ and $(3, 2)$.

$$\begin{array}{cc}(-4,-3) & (3,2)\\(x_1,y_1) & (x_2,y_2)\end{array}$$

$$Distance = \sqrt{(x_1 - x_2)^2 + (y_1 - y_2)^2}$$
$$= \sqrt{(-4 - 3)^2 + (-3 - 2)^2}$$
$$= \sqrt{(-7)^2 + (-5)^2}$$
$$= \sqrt{49 + 25}$$
$$= \sqrt{74} = 8.6$$

PRACTICE 28

Find the distance between each pair of points. Round your answer to the nearest tenth, if necessary.

1)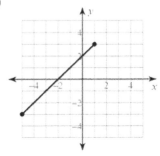

2)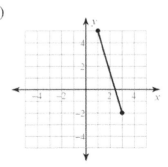

3)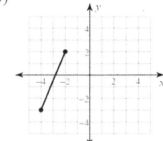

4)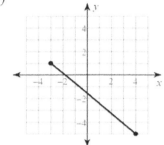

5) $(-7, 3), (-5, 0)$

6) $(4, 5), (6, -3)$

7) $(-6, -8), (5, -6)$

8) $(-2, 8), (2, 7)$

9) $(-0.5, 7.1), (-7.8, -5.8)$

10) $(1.2, 5.8), (-2.1, -0.2)$

11) (−6.6, 6.2), (4.4, −3.9)

12) (2.2, 4.3), (−0.6, 1.8)

13)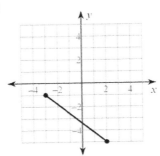

14)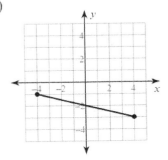

15) (−3, 0), (7, 1)

16) (4, −1), (−8, −8)

17) (4, −3), (−7, 7)

18) (−4, −8), (4, 5)

19) (0, −4), (8, 7)

20) (8, −8), (0, −7)

FINDING PERIMETER, AREA AND CIRCUMFERENCE

Perimeter is the distance around a shape measure in linear units such as inches, feet, centimeters, meters, etc. Circumference is the distance around a circle. When finding the perimeter of a shape such as a triangle, we just add up the lengths of the sides. There are some shortcut formulas though.

Rectangle: $P = 2l + 2w$, where l is the length and w is the width

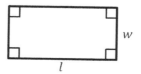

Square: $P = 4s$, where s is the side length (all sides of a square are the same)

Circle: $C = 2\pi r$, circumference is equal to two times the radius times the constant $\pi \approx 3.14$

The formula can also be written as $C = \pi d$ where d is the diameter.

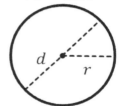

Area is the amount of space inside a shape measures in square units such as square inches, sq. feet, cm^2, m^2, etc. If you were remodeling a room, you would measure the perimeter to determine how much molding would go around the room, while you would use the area to determine how much paint you would need. Here we need more formulas:

Rectangle: $A = lw$

Square: $A = s^2$

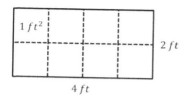

If we have a rectangle that is 2 ft by 4 ft then it would contain 8 square feet.

A parallelogram does not have sides that are perpendicular so we can't directly find the area from the sides of a parallelogram. But any parallelogram can be cut and put back together to form a rectangle. 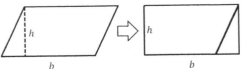 This helps us derive the formula, we use the perpendicular measurements of the base and height.

Parallelogram: $A = bh$

When we find the area of a parallelogram or a triangle, we must use the base and height. This can be somewhat confusing because the base is not necessarily the bottom side. The trick is to find the square telling us where the right angle is. The sides the square touches will be 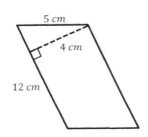 our base and height. In the picture on the right, the little square is on the dashed line of 4 and the line of 12. So, our base is 12 cm and the height is 4 cm. Therefore, the area would be 48 cm². When we calculate the perimeter, we would use the 5. $P = 5 + 12 + 5 + 12 = 34$ cm.

Any triangle is half a parallelogram so the formula for the area of a triangle is half that of 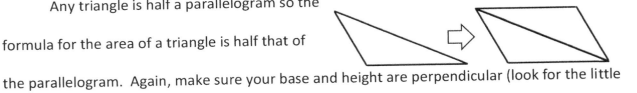 the parallelogram. Again, make sure your base and height are perpendicular (look for the little square).

Triangle: $A = \frac{1}{2}bh$

Find the perimeter and area of the triangle

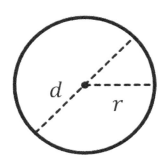

In the triangle to the right the square is on the outside. The height is the dashed side of length 4 on the outside. The base is 8. The $area = \frac{1}{2}(4)(8) = 16 \ m^2$. For the perimeter, we just add up all the sides. $Perimeter = 6 + 8 + 10 = 24 \ m$.

Circle: $A = \pi r^2$

The area formula for a circle uses the same numbers as the circumference formula: $2, \pi, r$. They are just in a different order. Area gives us squares so it makes sense to square the radius.

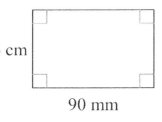

Ex. 1 Find the perimeter and area of the rectangle

First notice we have mixed units of centimeters and millimeters. We need to get these into the same units and it's best to use the bigger unit which will make our numbers we work with smaller, in this case we use cm. 90 mm = 9 cm. Now we can add up all the sides[27] to get the perimeter, or we can use the shortcut formula for a rectangle P = 2(6)+2(9) = 30 cm. To find the area we multiply the length of 9 cm by the width of 6 cm to get 54 cm².

---

[27] Opposite sides of a rectangle or any parallelogram are congruent.

Ex. 2 Find the perimeter and area of the parallelogram

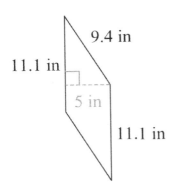

For the perimeter we need to know all four sides. Notice that the dashed line of 5 inches is not part of the outside of the parallelogram. The side on the bottom is not labeled but it's congruent to the side at the top. We add 9.4 + 11.1 + 9.4 + 11.1 = 41 inches. For the area we need to know the base and the height. Look for the square. It connects the dashed 5 inch line and the 11.1 inch side. We multiply these together to get 55.5 in².

Ex. 3 Find the perimeter and area of the triangle

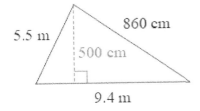

In this example, we have mixed units. We need to convert all of them into the same units. It is best to use the bigger units converting them all into meters. $500\ cm = 5\ m$ and $860\ cm = 8.6\ m$. The perimeter = 5.5 + 8.6 + 9.4 = 23.5 m. The area = ½(9.4)(5) = 23.5 m².

Ex. 4 Find the circumference and area of the circle.

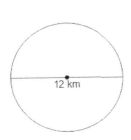

Here we are given the diameter of 12 km; the radius is half which is 6 km. The circumference $C = 2\pi r = 2\pi(6) = 37.699$ or rounded to nearest tenth we get 37.7 km. The area $A = \pi r^2 = \pi(6)^2 = 36\pi = 113.097$ The area is 113.1 km².

GEOMETRY

SHAPE	PERIMETER	AREA
Triangle	add up all the sides	$A = \dfrac{1}{2}bh$
Square	$P = 4s$	$A = s^2$
Rectangle	$P = 2l + 2w$	$A = lw$
Parallelogram	add up all the sides	$A = bh$
Circle	$C = 2\pi r$	$A = \pi r^2$

PRACTICE 29
Find the perimeter (circumference) and area of each.

1)
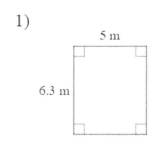

2)
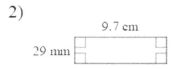

3)
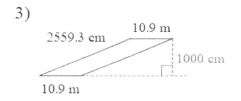

4)
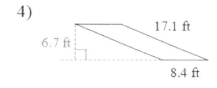

5)
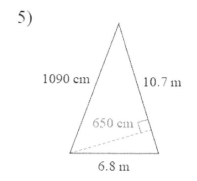

6)
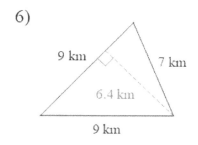

7)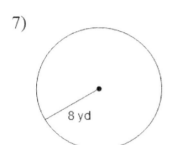

8)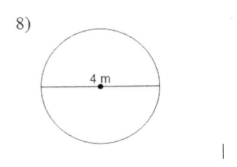

9)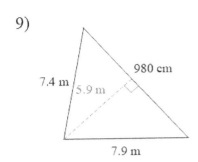

10)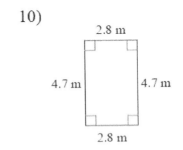

11)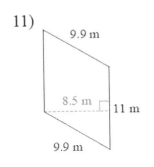

12)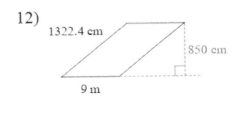

13)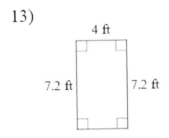

14)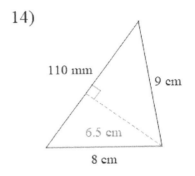

15)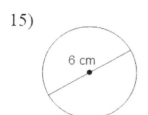

16)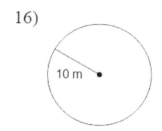

17)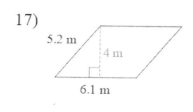

18)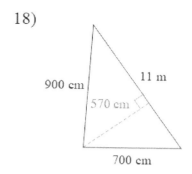

19)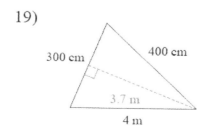

20)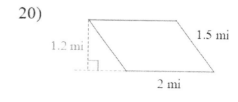

GEOMETRY

FINDING AREA AND PERIMETER OF COMPOSITE SHAPES

A composite or compound shape is one which is built from two or more of our basic shapes. Sometimes the shapes are "glued" or added together; sometimes one shape is cut out or subtracted from another. We will "break" the composite figure into its parts and either add or subtract the parts together.

<u>Ex. 1</u> Find the area and perimeter of the compound shape shown here:

First notice that it is made by adding a rectangle and semicircle. To find the area we find the area of the rectangle and add it to the area of half the circle. Fill in the missing sides. The top side is also 20 ft and the diameter of the semicircle is 8 ft.

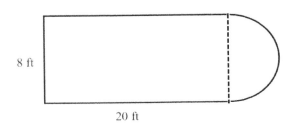

$$A_{total} = A_{rectangle} + A_{semicircle} = lw + \frac{1}{2}\pi r^2 = 20(8) + \frac{1}{2}\pi(4)^2 = 160 + 25.13 = 185.13$$

The total area is about 185.1 ft². For the perimeter, add up the lengths going around the outside, this includes half the circumference.

$P = 20 + 8 + 20 + \frac{1}{2}2\pi(4) = 20 + 8 + 20 + 12.57 = 60.57$ The total perimeter is 60.6 ft.

Ex. 2 Find the area and perimeter of the compound shape shown here:

Now notice that it is made by cutting out a triangle from a rectangle. To find the area we find the area of the rectangle and subtract the area of the triangle. Fill in the missing sides. The dashed lines at the bottom of the triangle will be 6 because the rectangle whole length is 20 and we need to subtract 7 from each end.

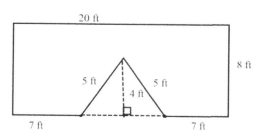

$A_{total} = A_{rectangle} - A_{triangle} = lw - \frac{1}{2}bh = 20(8) + \frac{1}{2}6(4) = 160 - 12 = 148$ The total area is 148 sq. ft.

For the perimeter make sure to add up all the sides on the outside (not the dashed lines, they are removed):

$P = 8 + 7 + 5 + 5 + 7 + 8 + 20 = 60$ The total perimeter is 60 ft.

Ex. 3 Find the area and perimeter of the compound shape shown here:

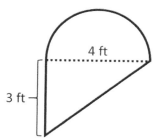

Here we have a composite shape made of a triangle and a semicircle. First, we find the area $\quad A_{total} = A_{triangle} + A_{semicircle} = \frac{1}{2}bh + \frac{1}{2}\pi r^2$

$= \frac{1}{2}(4)(3) + \frac{1}{2}\pi(2)^2 = 6 + 6.28 = 12.28\ ft^2$

To find the perimeter we need to know the missing side of the triangle. The Pythagorean Theorem can be used, $3^2 + 4^2 = 5^2$ to get a hypotenuse of 5. So, the perimeter will be

$5 + 3 + \frac{1}{2} circumference = 5 + 3 + \frac{1}{2}2(\pi)(4) = 8 + 4\pi = 20.57\ ft.$

PRACTICE 30
Find the perimeter and area of the composite figures

1.
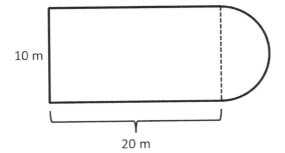

2.
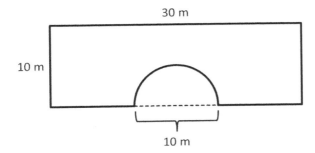

3. [28]
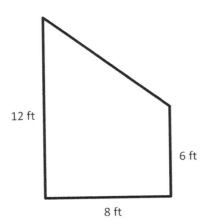

---
[28] You may recognize this shape as a trapezoid which has a formula for the area $A = 1/2(b_1+b_2)h$

CONGRUENT TRIANGLES

Proving triangles are congruent (the same) usually involves making sure we know for sure each of the three sides and each of the 3 angles matches up exactly to another triangle. But in fact, we don't need to know all 6 parts. It is enough if we know that 3 specific parts of one triangle match up to 3 specific parts of another triangle.

Side-Side-Side (SSS): If three sides of a triangle match up to 3 sides of another triangle, then the triangles are congruent. In the picture to the right, it appears we only have 2 pairs of sides marked congruent, but we also have the common side that forms both triangles for the third side.

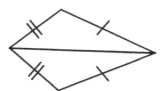

Side-Angle-Side (SAS): If two sides of a triangle are congruent to two sides of another triangle and the angles between each pair are congruent, then the triangles are congruent. In the picture to the right, it appears we only have 2 pairs of sides marked congruent, but we also have the vertical angle formed in the middle of the triangles for the angle.

Angle-Side-Angle (ASA): If two angles of a triangle are congruent to two angles of another triangle and the sides between each pair are congruent, then the triangles are congruent.

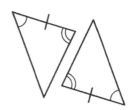

Angle-Angle-Side (AAS): If two angles of a triangle are congruent to two angles of another triangle and a side on the outside each pair are congruent, then the triangles are congruent.

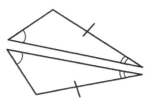

Four combinations of angles and sides, but not every combination works. Notice the ones that don't work for instance AAA (angle-angle-angle) doesn't work because we can take a triangle and enlarge or shrink it, the angles remain the same, but they are not congruent. We don't even need AAA, just AA. Two angles are enough because once we know two angles, we know the third. AA only works for similarity. The triangles are proportional to each other but not congruent.

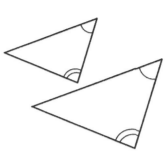

Also, SSA doesn't work because it's ASS backwards. There's no ASS in geometry. Here's a visual proof to demonstrate why it doesn't work. Obviously these two triangles are NOT CONGRUENT!

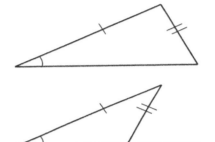

When doing these types of problems, keep in mind, you only "know" what is given to you by the markings on the diagram. If you have a common side, you can add a mark. If you have vertical angles, you can mark those angles congruent. Otherwise, you can't go by what appears to be the same.

<u>Ex. 1</u> Determine if the two triangles are congruent. If they are, state how you know.

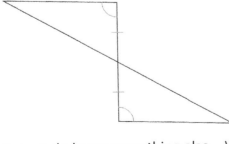

We are given a pair of congruent angles and a pair of congruent sides. As it is now, we don't have enough information.

But we do know something else... We can mark vertical angles congruent. Now we have enough information two pairs of congruent angles and a side that is sandwiched between the angles. Therefore, this is ASA.

Ex. 2 Determine if the two triangles are congruent. If they are, state how you know.

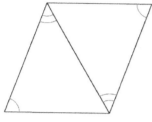

Here we have two angles marked. We can also mark the common side between the two triangles. Since the side is not between the angles, we have AAS

Ex. 3 Determine if the two triangles are congruent. If they are, state how you know.

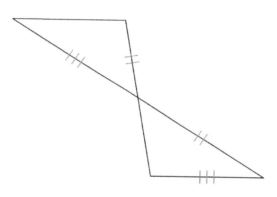

In this example we have two pairs of sides marked. We can also mark the vertical angles as congruent. It seems like it would be SAS, but if we examine closely notice that in the top triangle the congruent angle is between the two marked sides whereas in the bottom triangle it is not between the two marked sides. In the top triangle we have SAS, but in the bottom, we have ASS. Remember, there's no ASS in geometry, therefore these two triangles are not congruent.

PRACTICE 31

Determine if the two triangles are congruent. If they are, state how you know.

1)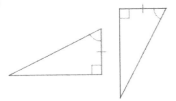

2)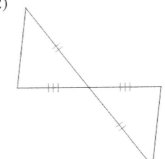

3)

4)

5)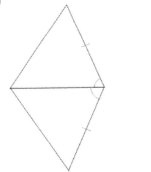

6)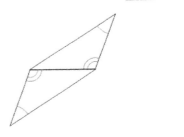

7)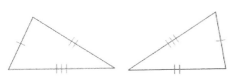

8)

9)

10)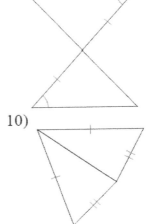

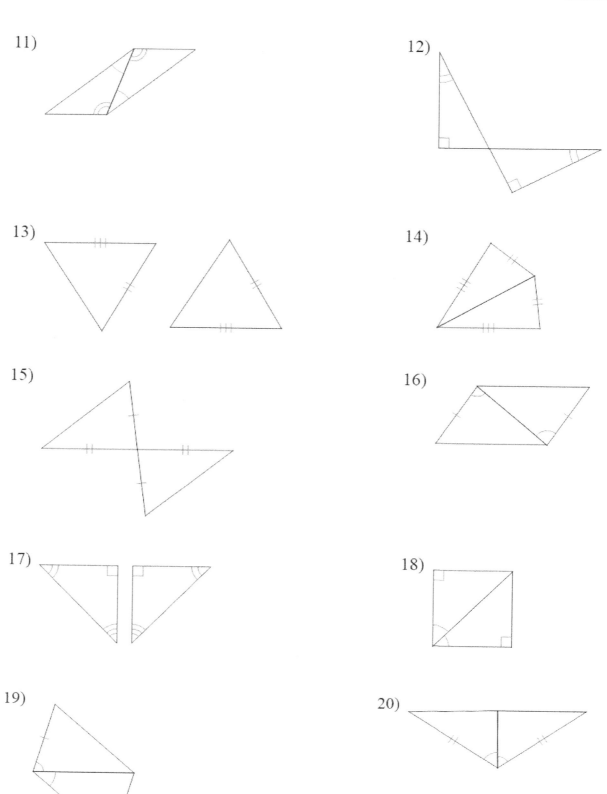

SIMILAR TRIANGLES

We mentioned similar triangles briefly. Now we will go into more details. Similar triangles have the same shape, same angle measures, but they are not the same size. However, the sides must all be in the same proportion. Just as with congruent triangles there are some shortcuts to determine if two triangles are similar.

Side-Side-Side (SSS): If three sides of a triangle are in the same proportion to 3 sides of another triangle, then the triangles are similar.

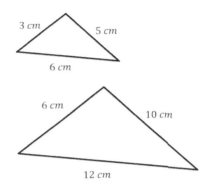

Side-Angle-Side (SAS): If two sides of a triangle are in proportion to two sides of another triangle, and the angle between is congruent, then the triangles are similar.

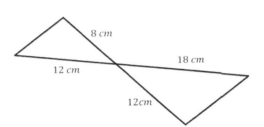

Angle-Angle (AA): If two angles of a triangle are congruent to two angles in another triangle, then the triangles are similar.

GEOMETRY

Ex. 1 State if the triangles are similar and if so, state the reason.

△EFG ~ △ERS

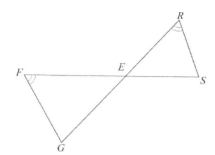

In this example we have only one pair of congruent angles marked. We can also mark the vertical angles congruent. These two triangles are similar by AA

Ex. 2 State if the triangles are similar and if so, state the reason.

△DEF ~ △DST

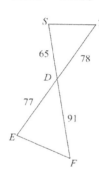

Here we are given the lengths of two pairs of sides and we can also mark the vertical angles congruent. It appears that we may have SAS similarity, but we need to check our proportions. Does $\frac{77}{65} = \frac{91}{78}$? We can see that they are not the same; 1.1846 does not equal 1.1666. Therefore, these triangles are NOT SIMILAR.

Ex. 3 State if the triangles are similar and if so, state the reason.

△KLM ~ △TSR

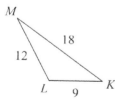

In this last example, we are given the lengths of three pairs of sides. We need to check that all three sides are proportional. Are these proportions all the same? $\frac{18}{12} = \frac{12}{8} = \frac{9}{6}$ They all reduce to $\frac{3}{2}$. These triangles are similar because of SSS.

PRACTICE 32

State if the triangles in each pair are similar. If so, state how you know they are similar.

1) △HGF ~ △UVW

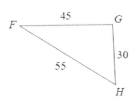

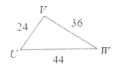

2) △BCD ~ △GFE

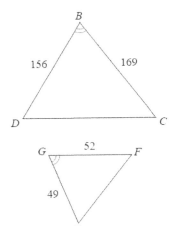

3) △EFG ~ △EVW

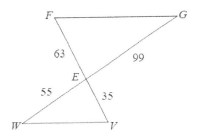

4) △FED ~ △LMN

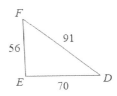

5) △RQP ~ △RLM

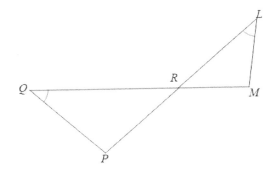

6) △VUT ~ △EFG

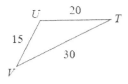

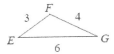

Page | 121

7)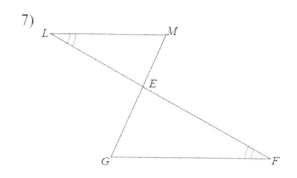

8) △NML ~ △CBA

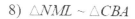
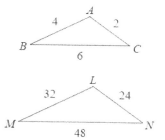

Find the missing length. The triangles in each pair are similar.

9) △HGF ~ △ABC

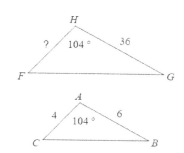

10)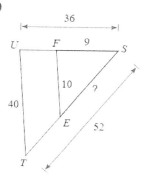

11) △ABC ~ △AEF

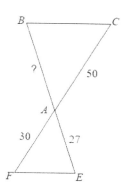

12) △STU ~ △CDE

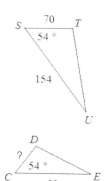

GEOMETRY

FINDING VOLUME

Volume is the amount of space a solid contains measured in cube units. Here's some common volume formulas:

Rectangular Prism	Cube	Sphere	Cylinder	Cone
$V = lwh$	$V = s^3$	$V = \frac{4}{3}\pi r^3$	$V = \pi r^2 h$	$V = \frac{1}{3}\pi r^2 h$

Ex. 1 Find the volume. Round answer to nearest hundredth.

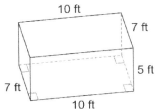

This is a rectangular prism. The formula for volume is length times width time height. In some problems, you may be given extra, unnecessary information. The length is 10 ft, the width is 7 ft and the height is 5 ft. Multiplying these together we get 350 ft³. Notice we have cubic units when measuring volume.

Ex. 2 Find the volume. Round answer to nearest hundredth.

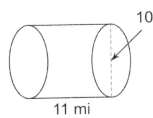

This is a cylinder so we must find the height and the radius. First find the radius which is half the distance across the circular base. We are given the diameter, whole distance across, is 10 mi so the radius is 5 mi. The height is the other dimension, 11 mi. Notice that the height for cylinders and cones does not need to be vertical. $V = \pi r^2 h = 3.14(5^2)(11) = 863.94 \, mi^2$. Using the calculator's pi button will give a more accurate answer.

PRACTICE 33

Find the volume of each figure. Round your answers to the nearest hundredth, if necessary.

1)
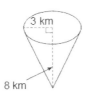

2)
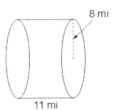

3)
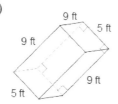

4)
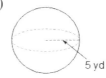

5)

6)
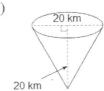

7)

8)

9)
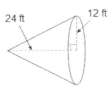

10)
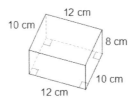

11)

12)

ALGEBRA

Algebra is the third branch of mathematics that we now explore in our intro to college math. Algebra is a mathematical branch that uses symbols to represent and manipulate unknown quantities to provide solutions to more complex problems than basic arithmetic alone can solve. A **variable** is a letter that represents a number, like x, y, n, etc. A **constant** is a known number, like $2, -1, \pi$, etc. A **coefficient** is a number that is being multiplied by a variable and written in front of the variable, like the 2 in $2x$. An algebraic **expression** is a combination of variables, numbers, and mathematical operations.

SIMPLIFYING EXPRESSIONS

A **term** is part of an algebraic expression separated by a plus sign. For example, $2x^2 + 3y + 7$ has three terms: $2x^2, 3y, 7$. Subtraction can be rewritten as addition, adding the opposite. So, we can write $5x^2 - 2xy + 3y - 7$ as $5x^2 + (-2xy) + 3y + (-7)$. We can see it has 4 terms: $5x^2, -2xy, 3y, -7$. **Like terms** have the same variables and exponents. $2x^2$ and $5x^2$ are like terms. $3x^2$ and $7x$ are *not* like terms. We can simplify an expression by combining (adding/subtracting) like terms. When we combine like terms we add (or subtract) the coefficients (numbers in front) and keep the variable part the same. The variable part is like a label that doesn't change when we combine like terms.

(2a + p + 3b) + (2a + 2p + 2b) = 4a + 3p + 5b

For example, in $5x - 2y + 3 + 8x + 6y - 9$ we can combine like terms. Underlining like terms: $5x - 2y + 3 + 8x + 6y - 9 = 13x + 4y - 6$

Ex. 1 Simplify $5x - 2y - 3 - 5x - 6y + 9$

Combine $5x$ and $-5x$ which comes out to $0x$ but we don't need to write it in our final answer. Then combine $-2y$ and $-6y$ which comes out to $-8y$. Finally combine -3 and 9 which comes out to 6. Put it all together to get: $-8y + 6$.

There are several properties that we can use to simplify expressions and you probably already use them without thinking. They are commutative, associative, and distributive properties and work with addition and multiplication.

NAME	DEFINITION	EXAMPLES
Commutative	You can change the order in addition and multiplication	$2 + 3 = 3 + 2$ $2 + x = x + 2$ $2 \cdot 3 = 3 \cdot 2$ $x \cdot 3 = 3x$
Associative	You can change the grouping (parentheses) in addition and multiplication	$(2 + 3) + 4 = 2 + (3 + 4)$ $(x + 3) + 4 = x + (3 + 4)$ $(2 \cdot 3)4 = 2(3 \cdot 4)$ $2(3x) = (2 \cdot 3)x$
Distributive	You can multiply a sum by multiplying each addend separately and then add the products.	$2(3 + 4) = 2 \cdot 3 + 2 \cdot 4$ $2(x + 4) = 2x + 2 \cdot 4$
Identity	When you add zero it doesn't change. When you multiply by one it doesn't change	$3 + 0 = 3$ $y + 0 = y$ $1 \cdot 5 = 5$ $1n = n$
Inverse	When you add the opposite, you get zero. When you multiply by the reciprocal you get 1.	$3 + -3 = 0$ $z - z = 0$ $\frac{1}{5} \cdot 5 = 1$ $\frac{1}{n} n = 1$

The most important of these for simplifying expressions is the distributive property.

ALGEBRA

Ex. 2 Distribute $3(4x - 7)$

$3(4x - 7) = 3 \cdot 4x - 3 \cdot 7 = 12x - 21$. The 3 gets distributed to both terms inside the parentheses. We cannot simplify $12x - 21$ because we don't have like terms. We don't know what x is. We simply have the expression 12x-21 for our final answer.

Ex. 3 Simplify by distributing and combining like terms: $5(2x - 4) - 3(2x - 5)$

$5(2x - 4) - 3(2x - 5) = 10x - 20 - 6x + 15$ Notice that we must distribute twice here. The second time we distribute a -3 since the negative goes with the term. Now we combine. $10x - 20 - 6x + 15 = 4x - 5$.

Ex. 4 Simplify by distributing and combining like terms: $5 + 2(3x - 4) - 6x$

$5 + 2(3x - 4) - 6x = 5 + 6x - 8 - 6x = -3$. Here we must keep in mind order of operations. Multiplying by 2 comes before adding 5. Only the 2 is distributed, the 5 must wait. After distributing we combine like terms. 6x and -6x cancel out, so all we are left with is -3.

Ex. 5 Simplify by distributing and combining like terms: $2 - 4(7 + 3n) - (8n - 8) - n$

$2 - 4(7 + 3n) - (8n - 8) - n = 2 - 28 - 12n - 8n + 8 - n$ We first distribute twice. Now we combine like terms: $2 - 28 - 12n - 8n + 8 - n = -18 - 21n$ We could also write this as $-21n - 18$

PRACTICE 34

Simplify each expression.

1) $b - 6 + 3 + 10b$

2) $-2m + 5 + 1 + 4m$

3) $n + 5 + 2 + 5n$

4) $7x - 9 + 3 - 7x$

5) $-6(p + 7)$

6) $3(a + 10)$

7) $3(5p - 1)$

8) $-7(2b - 4)$

9) $-6m + 4(4 + 2m)$

10) $-7 + 2(-10x + 4)$

11) $-6(2x + 6) + 5$

12) $-10x + 9(x + 5)$

13) $-7(-p - 3) - 3(3p + 8)$

14) $8(2 - 3n) - 7(1 - 6n)$

15) $5(x - 1) + 2(4x + 4)$

16) $-3(x - 10) + 3(1 + 4x)$

17) $-10m - 5(1 - 6m)$

18) $2k + 10(8 - 6k)$

19) $-2(-1 + 10x) + 4(2x - 5)$

20) $-10(1 - 5n) - (n - 8)$

ALGEBRA

EVALUATING EXPRESSIONS AND USING ORDER OF OPERATIONS

Evaluating expressions, means that we substitute or plug in given values for a variable or variables. Do the operations and we end up with a number for our answer.

Ex. 1 Evaluate $m^2 + p - (m - q)$; use $m = -5, p = 4, q = 6$

Rewrite the expression, taking out the variables and putting parentheses in their place.

$(\ )^2 + (\ ) - ((\ ) - (\ ))$ Now fill in the gaps. Substitute the numbers in the corresponding spaces. Follow PEMDAS: $(-5)^2 + (4) - ((-5) - (6)) = (-5)^2 + (4) - (-11)$

$= 25 + (4) - (-11) = 29 - (-11) = 29 + 11 = 40$

Ex. 2 Evaluate $-\dfrac{xz}{|yz|} - zy$; use $x = 4, y = 2, z = -3$

$-\dfrac{(4)(-3)}{|(2)(-3)|} - (-3)(2) = -\dfrac{(4)(-3)}{|-6|} - (-3)(2) = -\dfrac{-12}{6} - (-3)(2) = -(-2) - (-3)(2)$

$= 2 - (-6) = 2 + 6 = 8$

PRACTICE 35

Evaluate each using the values given.

1) $y + z(y + z) - y$; use $y = -2$, and $z = 4$

2) $z + x^2 - (z + 3)$; use $x = -1$, and $z = -4$

3) $z(z - (x + 2) - z)$; use $x = -5$, and $z = 2$

4) $2(x - y)(y + y)$; use $x = 6$, and $y = 1$

5) $\dfrac{m}{|m|}(p^2 - m)$; use $m = -6$, and $p = -5$

6) $x - (-4x - (y + y))$; use $x = -4$, and $y = -3$

7) $z(z - 2) + y^3$; use $y = 2$, and $z = -6$

8) $z(z^2 - (-5 - x))$; use $x = 6$, and $z = -2$

SOLVING ONE-STEP AND TWO-STEP EQUATIONS

The simplest type of equation to solve is a one-step equation. We just undo what has been done to the equation to get the variable by itself. We will be focusing on linear equations in this course. Linear equations are equations that have one variable, like x, and it not raised to any higher exponent. Most linear equations will have one solution, although there are some that have no solution or infinite solutions. For example, $2x + 5 = 11$ is a linear equation with one solution, namely x = 3. This equation: $x + 1 = x$ has no solution since there's no number that when you add one to itself you get that number. Here's one with infinite solutions: $2(x + 3) = 2x + 6$. It has infinite solutions because if we distribute on the left side, we get exactly the same expression as on the right and if we take any number add three then multiply all of that by 2 it will equal the number times 2 plus 6. We will start from the easiest type of equations and work towards more difficult equations.

Ex. 1 Solve: $x + 7 = 15$

When solving an equation, we usually write our work this way.

$$\begin{aligned} x + 7 &= 15 \\ -7 &= -7 \\ \hline x &= 8 \end{aligned}$$

Whatever we do to one side, we must do the other side of the equation to keep it balanced. To undo adding 7 to x, we subtracted 7 from both sides to get x = 8.

Ex. 2 Solve: $2x = 14$. In this equation x is multiplied by 2, so we do the "opposite" operation and divide both sides by two. We get $x = 7$.

Ex. 3 Solve: $\frac{2}{3}x = 8$. In this equation x is multiplied by $\frac{2}{3}$, so we do the "opposite" operation and divide both sides by $\frac{2}{3}$. Dividing by a fraction is the same as multiplying by the reciprocal. $\frac{3}{2} \cdot \frac{2}{3}x = \frac{3}{2} \cdot 8$ We get $x = 12$.

Two-step equations means that two operations were performed to the variable and we have to undo the steps in the opposite order.

Ex. 4 Solve $2x + 5 = 21$.

What is happening to the x on the left side to get the number on the right? We take x and multiply by 2 then we add 5. To solve this, we must go in reverse order, doing the opposite of each operation. The last thing done was adding 5, so this is the first thing to undo by subtracting 5. $2x = 16$ Then the opposite of multiplying by 2 is dividing by 2. We get $x = 8$. You can always check your answer by plugging it back into the original problem. Does $2(8) + 5 = 21$? Yes, it works, so we know we have solved correctly.

Ex. 5 Solve $\frac{x-7}{3} = 4$.

One way to solve this problem is again thinking about what was done to x on the left side. First, we subtract 7, then we divide by 3. Going in reverse and doing the opposite we multiply by 3. $x - 7 = 12$. Then add 7 to both sides, $x = 19$. Checking our solution: $\frac{19-7}{3} = 4$ does work.

Ex. 6 Solve $\frac{4}{3}m + 1\frac{2}{5} = \frac{97}{30}$

This equation has some fractions to deal with. Here the variable is *m*. You didn't think the variable always had to be an *x*, did you? *X* is just usually our favorite variable to use. The variable *m* is being multiplied by $\frac{4}{3}$, then we add $1\frac{2}{5}$ to it. To get the *m* by itself we have to go backwards by subtracting $1\frac{2}{5}$ and dividing by $\frac{4}{3}$. To subtract fractions remember we need a common denominator and it usually much easier to use improper fractions when solving equations. $1\frac{2}{5} = \frac{7}{5} = \frac{42}{30}$. Now subtract from both sides $\frac{4}{3}m = \frac{97}{30} - \frac{42}{30}$. This gives us $\frac{4}{3}m = \frac{55}{30}$. Dividing by four thirds is the same as multiplying by the reciprocal three fourths. $\frac{3}{4}\frac{4}{3}m = \frac{3}{4}\frac{55}{30}$. Reducing the right side gives $\frac{3}{4}\frac{55}{30} = \frac{1}{4}\frac{55}{10} = \frac{1}{4}\frac{11}{2} = \frac{11}{8}$.

PRACTICE 36

Solve each equation.

1) $x - 10 = 0$

2) $3m = -42$

3) $22 = r + 5$

4) $9 = \dfrac{p}{8}$

5) $-57 = 3x$

6) $9 + x = 11$

7) $16 = 6n + 4$

8) $\dfrac{-2 + v}{4} = -2$

9) $-8 = \dfrac{p}{7} - 6$

10) $-4(a + 6) = 24$

11) $\dfrac{94}{3} = -2 + \dfrac{5}{3}x$

12) $\dfrac{1}{3}n + \dfrac{1}{8} = \dfrac{13}{40}$

13) $\dfrac{3}{2} = \dfrac{1}{3}\left(-\dfrac{3}{2} + v\right)$

14) $-\dfrac{10}{7}p - \dfrac{3}{5} = \dfrac{22}{5}$

SOLVING MULTISTEP EQUATIONS

Now we get to more complicated equations with multiple steps. These require that we use all the algebra tools and properties that we have discussed so far. Here's the steps that we go through to solve a multistep equation. Some steps may be skipped at times.

1. Simplify the left side and right side by combining like terms and using distributive property
2. Move all the variable terms to one side (usually the left side) by adding or subtracting.
3. Move all the constant terms to the other side (usually the right side) by adding or subtracting
4. Divide by the coefficient of the variable
5. Check your answer by plugging it back into the original equation

<u>Ex. 1</u> Solve $-3 + 3(-6x + 5) = -2x - 36$

Distribute: $-3 + -18x + 15 = -2x - 36$

Combine like terms: $-18x + 12 = -2x - 36$

Add 2x to both sides: $-16x + 12 = -36$

Subtract 12 from both sides: $-16x = -48$

Divide both sides by -16: $x = 3$

Check the solution: $-3 + 3(-6(3) + 5) = -2(3) - 36$ Both sides give -42 so it works.

ALGEBRA

Problems with fractions can be a little tricky. We will try to make them easier by multiplying each term by the LCD (least common denominator) of each denominator used.

Ex. 2 Solve $\frac{99}{40} - \frac{2}{5}x = -\frac{1}{4}x + \frac{3}{2}x$

The LCD of 40, 5, 4, and 2 is 40. Multiply each term by 40. Remember to make it easier by canceling. $\frac{40}{1}\frac{99}{40} - \frac{40}{1}\frac{2}{5}x = -\frac{40}{1}\frac{1}{4}x + \frac{40}{1}\frac{3}{2}x$ becomes $\frac{1}{1}\frac{99}{1} - \frac{8}{1}\frac{2}{1}x = -\frac{10}{1}\frac{1}{1}x + \frac{20}{1}\frac{3}{1}x$ then $99 - 16x = -10x + 60x$ Now combine like terms on the right $99 - 16x = 50x$ Since all the right side has just variable terms, we can move the other variable term to the right side by adding 16x to both sides $99 = 66x$ Divide by the coefficient of 66 to get $\frac{99}{66} = x$ which reduces to $x = \frac{3}{2}$

Dear Algebra,

Please stop asking to find your x. She's never coming back and don't ask y.

The Romans never found algebra very challenging because X was always 10.

PRACTICE 37

Solve each equation.

1) $-1 + 2x = 7 + 4x$

2) $4p + 8 - p = p + 8 + 7 - 5$

3) $5(1 - 3n) = -100$

4) $7 - 7(m - 6) = 98$

5) $-6(4 + 4x) + 4x = 36 - 8x$

6) $-8 - 7(2 - 5n) = 8n + 32$

7) $8(x - 2) = 2x + 2(2x - 2)$

8) $1 - 8n - 7 = -5(8 - 2n) - 5(5n - 4)$

9) $2r + \dfrac{5}{4}r = \dfrac{65}{8}$

10) $n + \dfrac{3}{2} - \dfrac{3}{2}n = -\dfrac{1}{8}$

11) $\dfrac{8}{3}x + \dfrac{2}{5} + \dfrac{23}{5}x + \dfrac{11}{45} = 6x + \dfrac{7}{3}$

12) $-\dfrac{17}{24} - \dfrac{11}{6}x = \dfrac{1}{3}x + \dfrac{9}{2} + 2x$

13) $-6(6b - 8) = -3b + 3(2b + 3)$

14) $5(a + 6) = -4(a - 3)$

15) $-6n - 20 = -2(n - 4)$

16) $-5x + 13 = 6(-2 - 5x)$

17) $-12 - 4n = -6n - 4$

18) $5 + 3n = -10 + 3n + 3n$

19) $-5(-7 + 5x) = 185$

20) $8(3x + 8) = 232$

21) $-\dfrac{3}{2}n - \dfrac{37}{4} = -\dfrac{4}{3}n + 6n$

22) $\dfrac{41}{16} - \dfrac{1}{2}r = -\dfrac{7}{4}r + 1$

ALGEBRA

SOLVING INEQUALITIES

Inequalities include statements that include $<$ less than, $\leq$ less than or equal to, $>$ greater than, $\geq$ greater than or equal to. Inequalities will have many solutions. Suppose we have $x < 4$. There's an infinite amount of numbers less than 4, such as 3, -1, 2.5, π, etc. In order to visualize solutions, we will often graph our solutions as well on a number line. If we have a strict inequality (not equal to) then we use an open circle. If we have equality, we use a closed or filled-in circle. Here's some examples of inequalities and graphs.

$x < 4$	$x \leq 4$	$x > 4$	$x \geq 4$
←—○—→ 2 3 4 5 6	←—●—→ 2 3 4 5 6	←—○—→ 2 3 4 5 6	←—●—→ 2 3 4 5 6

Solving inequalities will be very similar to solving equations with two exceptions.

1. Whenever you multiply/divide by a negative you need to flip the inequality sign

    Ex. $-3x < 12$ Divide by -3 and we flip the inequality sign to get $x > -4$

2. If you switch the sides of an inequality, then the inequality sign will flip as well.

    Ex. $-4 \leq x$ is the same as Ex. $x \geq -4$

Ex. 1 Solve and Graph

$$-7n - 21 > 7(3 + 5n)$$

←——+——+——+——+——+——+——+——+——+——+——→
-8 -7 -6 -5 -4 -3 -2 -1 0 1 2

$-7n - 21 > 7(3 + 5n)$

$-7n - 21 > 21 + 35n$ Distribute on the right side

$-42n - 21 > 21$ Subtract $35n$ from both sides

$-42n > 42$ Add 21 to both sides

Page | 138

$n < -1$ Divide both sides by -42. Since we divide by a negative, we flip the inequality sign. We put an open circle at the negative one and shade to the left.

Ex. 2 Solve and Graph $-(a - 4) - 5a > -22 + 7a$

$-(a - 4) - 5a \geq -22 + 7a$

$-a + 4 - 5a \geq -22 + 7a$ Distribute the negative on the left side of the equation

$-6a + 4 \geq -22 + 7a$ Combine like terms of -a and -5a to give -6a

$4 \geq -22 + 13a$ Add 6a to both sides. Notice I said usually we want to move the variables to the left but it's not necessary. As long as we do the same operation to both sides, we'll be good.

$26 \geq 13a$ Add 22 to both sides

$2 \geq a$ Divide both sides by 13. Here we have the variable by itself, but it's a little more difficult to think about what this solution means. It much easier to rewrite the solution with the variable on the left side. To do this we switch signs and flip the inequality symbol.

$a \leq 2$ This is easier to read as *a is less than or equal to 2*. We put a solid, filled in circle on 2 and shade to the left.

PRACTICE 38

Solve each inequality and graph its solution.

1) $-11 > a - 6$

2) $\dfrac{n}{2} \leq -8$

3) $5(k + 6) > -15$

4) $\dfrac{2 + a}{2} \leq -7$

5) $11 + 2m + 8m < 8 + 7m$

6) $12 - 8m + 7 - 3m > 7 - 8m$

7) $-3b + 5 + 7 < 14 - b$

8) $6a + 2a \leq 6a + 12$

9) $2(v+5) > -25 - 3v$

10) $8(3n-2) > -16 - 8n$

11) $6(1+8n) \geq 6(1-n)$

12) $5x + 6x > 7(1+x) - 7(1+8x)$

13) $x - \dfrac{5}{7} > \dfrac{1}{3}x - \dfrac{22}{21}$

14) $\dfrac{8}{7}b + \dfrac{1}{4} < -\dfrac{37}{28} + \dfrac{5}{2}b + b$

15) $7(x+2) < -5x - (4-3x)$

16) $4n + 2(3n+1) \geq 2(n-7)$

17) $-26 - 6k > -8(1 + 3k)$

18) $-3p + 34 \leq -8(6p + 7)$

19) $-14 + 6n > -3n + 2n$

20) $8 + 2n < n + 6 - 6$

21) $14 < 5x + 2x$

22) $-4n - 7n < 0$

23) $\dfrac{16}{7}v + \dfrac{81}{28} < v + \dfrac{3}{4}$

24) $\dfrac{1}{28} + \dfrac{8}{7}x + \dfrac{3}{4}x \geq \dfrac{1}{2}x + \dfrac{1}{2}$

FINDING SLOPE AND RATE OF CHANGE

Graphs are used for many different purposes. Graphs may represent the growth of money over time, distance traveled over time, number of jobs over time, etc. We would to quantify the amount of change that happens. **Slope** measures the **rate of change** by the steepness of the line.

Here's a graph that shows the national supply and demand projections for registered nurses from 2000 to 2020. What can we conclude from this graph? Demand is

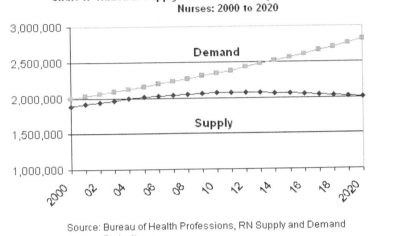

Chart 1: National Supply and Demand Projections for FTE Registered Nurses: 2000 to 2020

Source: Bureau of Health Professions, RN Supply and Demand Projections

increasing; it has a positive rate of change over time. The rate actually goes up around 2012, where it starts to go up faster. Supply starts increasing then reaches a point around 2010 where the rate of change is zero, and then the rate of change decreases.

The basic definition of slope is $m = \dfrac{rise}{run} = \dfrac{change\ in\ y}{change\ in\ x}$

Slope is "read" from left to right. So, if we are going uphill from left to right, we have positive slope, whereas going downhill is negative slope, a flat horizontal line has zero slope, and a vertical line has undefined slope. We never use the phrase "no slope."

ALGEBRA

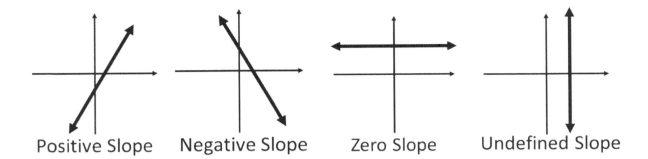

Positive Slope   Negative Slope   Zero Slope   Undefined Slope

Slope will be a number either a whole number, or a reduced fraction (may use reduced improper fractions). We will not use mixed numbers, percents, or decimals for slope. The higher number, the steeper the slope becomes.

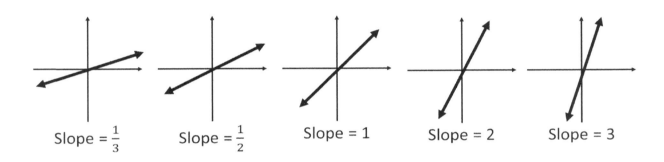

Slope = $\frac{1}{3}$ Slope = $\frac{1}{2}$ Slope = 1   Slope = 2   Slope = 3

FINDING SLOPE FROM A GRAPH

When finding slope from a graph, we need to find or identify two "nice" points, points that are on the corners of the boxes in our coordinate grid. We then make a triangle. The rise will be the distance of the vertical line, while the run will be the distance of the horizontal line.

Ex. 1 Find the slope from the graph

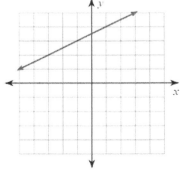

Page | 144

The first thing to do is to find some nice points on this line that fall on the corners. The nice points are (-3,2), (-1, 2), ((1,4). We only need 2 points, but it helps to find more to make sure you see the pattern continues. As we go from one point to another how far do we go up? How far do we go over? We can make a triangle and find the lengths of the sides. If we start from the point (-3, 2) we would have to go over 2 (run = 2) and up 5 (rise = 1). Therefore the slope is $m = \frac{1}{2}$

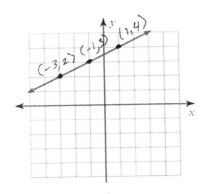

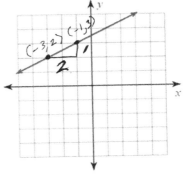

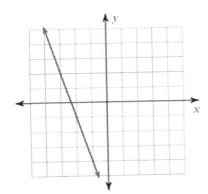

Ex. 2 Find the slope from the graph. Right away we know that the slope is going to be negative since it's going downhill. We pick out two points, (-3, 2) and (-1, -4). If we start at (-3, 2) we have to go down 6 (rise = -6) and we have to go right 2 (run = 2). Slope $m = \frac{-6}{2} = -3$. The slope is negative 3. We could also switch the order of the points and start at (-1, -4) and got to (-3, 2). Then, we go up 6 (rise = 6) and we go left 2 (run = -2), which gives $m = \frac{6}{-2} = -3$. It doesn't matter which point we start from; the slope will be the same.

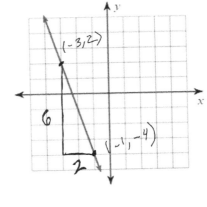

ALGEBRA

FINDING SLOPE FROM TWO POINTS

If we are given the coordinates of the two points, then we can use the slope formula to find the slope between two points.

$$m = \frac{y_2 - y_1}{x_2 - x_1}$$

Ex. 1 Find the slope between $(-3, 7)$ and $(17, 19)$. Plug into the formula $m = \frac{19-7}{17--3} = \frac{12}{20} = \frac{3}{5}$

Ex. 2 Find the slope between $(-3, 5)$ and $(-3, 10)$. Plug into the formula $m = \frac{10-5}{-3--3} = \frac{5}{0} =$ undefined

PRACTICE 39

Find the slope of each line.

1)

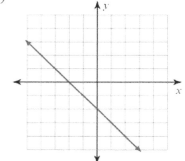

2)

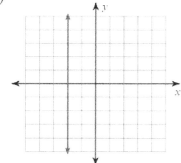

3)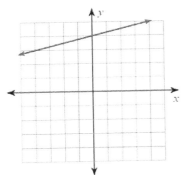

4)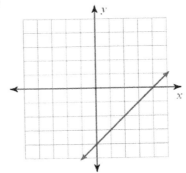

5)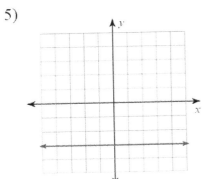

6)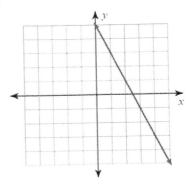

7)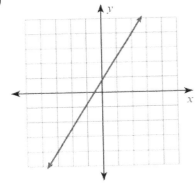

8)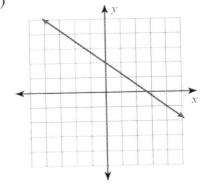

Find the slope of the line through each pair of points.

9) $(8, 9), (8, 16)$

10) $(-14, 0), (-8, 15)$

11) $(-13, -20), (4, -3)$

12) $(-17, 6), (-19, 5)$

13) $(11, 3), (5, 17)$

14) $(2, 19), (1, 20)$

15) $(18, 13), (14, -5)$

16) $(-12, -6), (0, -6)$

LINEAR EQUATIONS

We have looked at finding slope from the graph of a line. We can write an equation with two variables that can be represented by a linear graph. The equations can be written in **slope-intercept form**: $y = mx + b$ where m represents the slope and b represents the y-intercept (where the line crosses the y-axis). In this equation, x and y are variables. For every number that we substitute into x, we get a specific y value.

GRAPHING LINEAR EQUATIONS

We will go through a few ways of graphing linear equations: 1) making a table, 2) using the y-intercept and slope, 3) using a graphing calculator.

ALGEBRA

Ex. 1 Graph the line $y = 2x + 3$

x	y
-2	$2(-2) + 3 = -1$
-1	$2(-1) + 3 = 1$
0	$2(0) + 3 = 3$
1	$2(1) + 3 = 5$
2	$2(2) + 3 = 7$

We select a few values to plug in for x. It is recommended to usually plug in 0, some negative numbers, and some positive numbers. X is the independent variable; we are free to choose which number to plug in. Y is the dependent variable; the value for y depends on what we plug in for x. From this table we use the corresponding x's and y's to get a set of coordinate points to graph.

$(-2, -1), (-1, 1), (0, 3), (1, 5),$ and $(2, 7)$. Now plot the points and connect the points.

Same example, but intercept/slope method: Plot the y-intercept first. Then use the slope for directions to the next point, plot and repeat. Connect the points.

$$y = 2x + 3$$

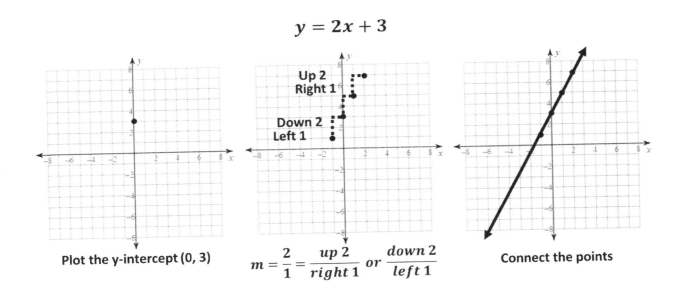

Plot the y-intercept (0, 3) $m = \dfrac{2}{1} = \dfrac{up\ 2}{right\ 1}$ or $\dfrac{down\ 2}{left\ 1}$ Connect the points

Ex. 2 $3x + 2y = 6$

In this example we have an equation in standard form. When an equation is in standard form, we need to solve for y and write it in slope-intercept form ($y = mx + b$) to graph it easily. The first step is to add/subtract the x term to take the x's to the other side. In $3x + 2y = 6$, we would subtract 3x from both sides to get $2y = -3x + 6$. Notice that we can't combine the -3x and the 6 because they are not like terms. We also change the order and write the -3x first.

Next, divide every term on both sides by 2: $\frac{2y}{2} = \frac{-3x}{2} + \frac{6}{2}$

When we simplify, we get: $y = -\frac{3}{2}x + 3$. In slope-intercept form we know the y-intercept is 3 and the slope is $-\frac{3}{2}$. Plot the point at (0, 3). Then from that point go down 3 and right 2, down 3 right 2. Connect the dots and we have our graph.

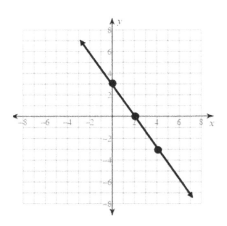

Ex. 3 $6x - 3y = 0$

Solve for y gives $3y = 6x$ which becomes $y = 2x$. This equation has no constant, so what is the y-intercept? The y-intercept is (0, 0). The slope is 2 which tells us to go up 2, to the right 1.

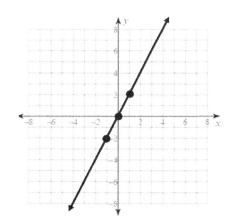

GRAPHING CALCULATOR HELP: GRAPHING LINES

Here we will use a TI-84 calculator to help graph a line. First press the [Y=] key at the top left of the calculator. You will see the y= window. Here we will type in our equation.

Type in 2x + 3 using the [X,T,θ,n] key for the x variable. Then hit the [GRAPH] button near the top right. You can see what the graph looks like. If it doesn't look like this try [ZOOM] [6]. You can bring up the table using [2ND] [GRAPH]. Use arrow keys move navigate.

SPECIAL CASES

Horizontal lines have equations of the form $y = b$, where b is a number (y-intercept). Vertical lines have equations of the form $x = a$, where a is a number (x-intercept). Here's a way to remember: **HOY VUX**

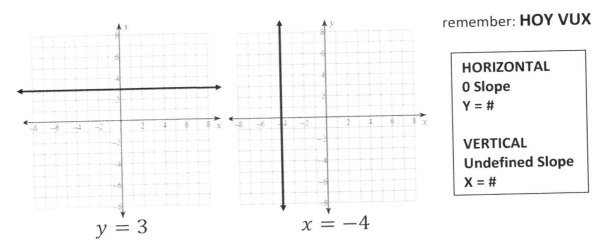

$y = 3$ $x = -4$

HORIZONTAL
0 Slope
Y = #

VERTICAL
Undefined Slope
X = #

PRACTICE 40

Sketch the graph of each line.

1) $y = -x - 2$

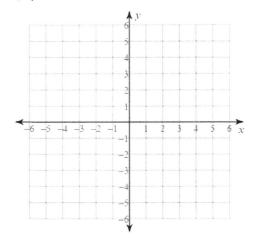

2) $y = \dfrac{3}{4}x - 1$

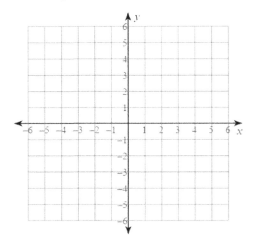

3) $y = \dfrac{2}{3}x + 3$

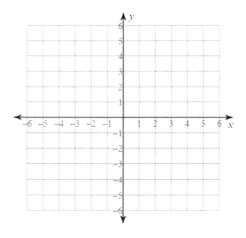

4) $y = 4x - 1$

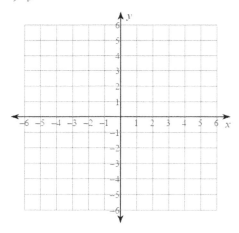

5) $y = 5x$

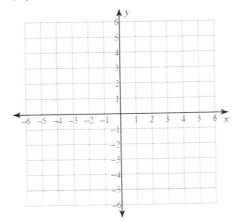

6) $y = 5$

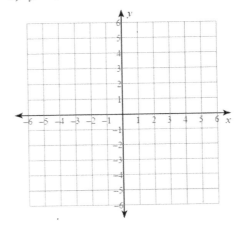

7) $y = -\dfrac{3}{4}x - 5$

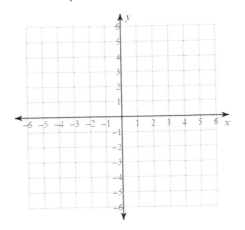

8) $y = x + 3$

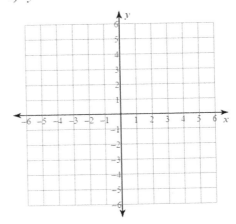

9) $x + 2y = -6$

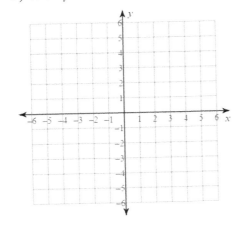

10) $x - y = 2$

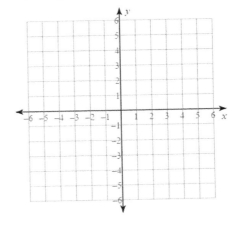

11) $x + y = 2$

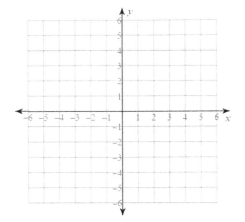

12) $5x + 4y = 20$

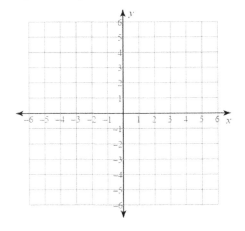

13) $x + 5y = 25$

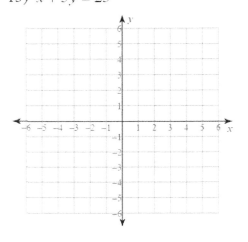

14) $x - y = -1$

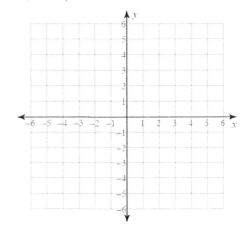

ALGEBRA

SOLVING SYSTEMS OF EQUATIONS

A **system of equations** is two or more equations in 2 (or more variables) to be solved simultaneously. We try to find a point(s) that work in both equations at the same time. We will solve a system of two equations with two variables. There are several ways to solve systems of equations: graphing, substitution, and elimination. We will focus on solving systems by graphing. First, we need to solve each equation for y. Then graph each equation on the same coordinate plane. Find the point(s) of intersection. Since we are graphing 2 lines, we have 3 different cases:

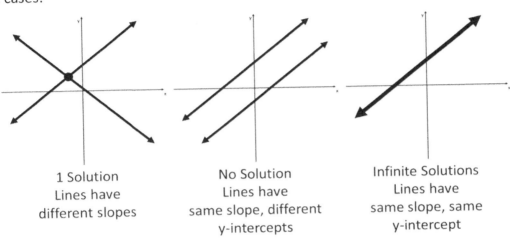

1 Solution
Lines have
different slopes

No Solution
Lines have
same slope, different
y-intercepts

Infinite Solutions
Lines have
same slope, same
y-intercept

Ex. 1 Solve $\begin{array}{l} y = -\frac{2}{3}x + 1 \\ y = x - 4 \end{array}$

The first equation has a y-intercept at 1, so we plot (0, 1). Then we go down 2 and right 3 (or up 2, left 3). Plot as many points as you can on the system, because your solution is probably going to be one of those points! Graph the other

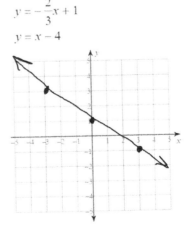

line which starts at (0, -4) and then goes up 1 right 1. Again, plot as many points as possible. Eventually you will come across a point that is on both lines. That is our solution. For this example, our solution is (3, -1). In this course and book, solutions will be integer solutions, so they will always fall on lattice points (corners of the boxes). No fractions or decimals.

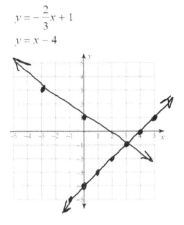

Ex. 2 Solve $\begin{array}{l} y = -\frac{1}{2}x + 1 \\ y = -\frac{1}{2}x - 4 \end{array}$

Here we have a system where both lines have the same slope. Since these lines will never intersect, we write "No Solution" for our answer. Even if there is no solution it is important to write "No Solution," Don't just leave the question blank.

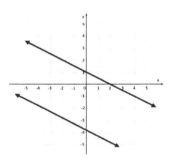

Ex. 3 Solve $\begin{array}{l} x + y = -3 \\ 3x - 4y = -16 \end{array}$

Here both equations are in standard form. So, we must solve each for y. Take the first equation $x + y = -3$ and subtract x from both sides to get $y = -x - 3$. Take the second equation $3x - 4y = -16$ and subtract 3x from both sides to get $-4y = -3x - 16$. Then we divide by -4 to get $y = \frac{3}{4}x + 4$. Now graph both and find the intersection.

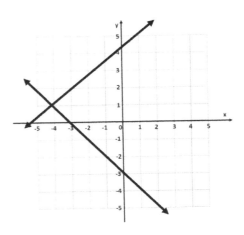

The solution is (-4, 1)

CALCULATOR HELP: SYSTEMS

We can use the graphing calculator to solve systems of equations by graphing. Both equations must be solved for y to put into the calculator. Press [Y=] (STAT PLOT F1) key and type the first equation into y₁ and the second equation into y₂. Type in $-x - 3$. Make sure to use [(-)] (ANS ?) at the bottom of the calculator for negative signs in front of the x. Type [X,T,θ,n] for the x. Then hit [GRAPH] (TABLE F5) You can see what the graph looks like. Zoom standard if needed [ZOOM] (FORMAT F3) [6] (L6 V). We can find the intersection point by going to [2ND] [TRACE] (CALC F4) then go down to 5: intersect then press [ENTER] (ENTRY SOLVE) three times. You can bring up the table using [2ND] [GRAPH] (TABLE F5). Use arrow keys move navigate.

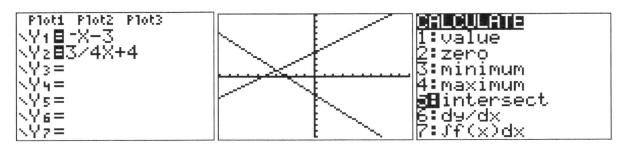
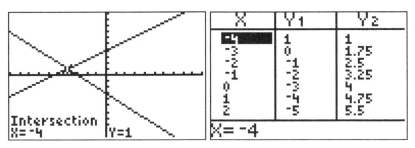

PRACTICE 41

Solve each system by graphing.

1) $y = -\dfrac{3}{2}x + 4$
   $y = -\dfrac{1}{4}x - 1$

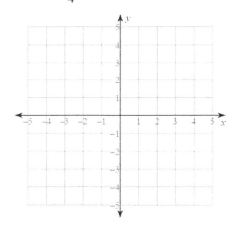

2) $y = -x + 1$
   $y = x - 3$

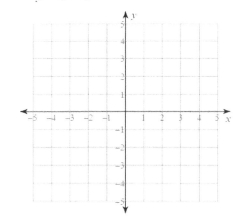

3) $y = -3$
   $y = -\frac{5}{2}x + 2$

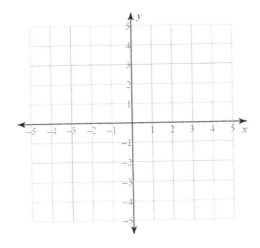

4) $y = \frac{3}{2}x - 3$
   $y = \frac{1}{2}x + 1$

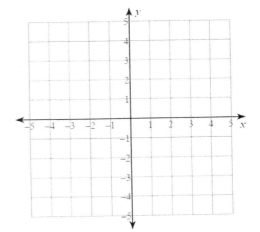

5) $y = -\frac{3}{2}x + 4$
   $y = 2x - 3$

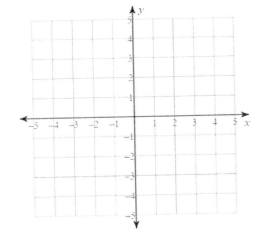

6) $y = -x + 4$
   $y = \frac{1}{2}x + 1$

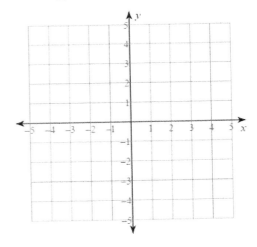

7) $y = -x - 2$
   $y = -4x + 4$

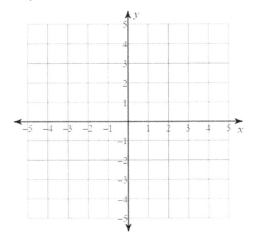

8) $y = \dfrac{1}{2}x - 2$
   $y = 3x + 3$

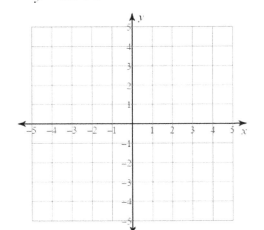

9) $3x - y = -4$
   $5x + y = -4$

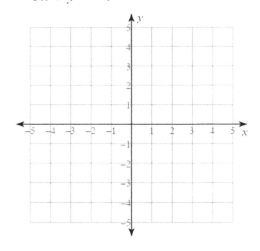

10) $x + 2y = 2$
    $x + 2y = 6$

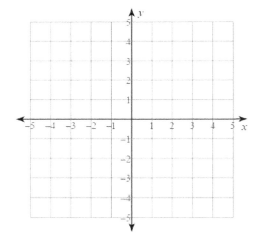

11) $x = 4$
$x + 2y = -2$

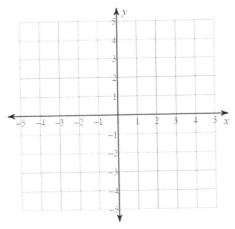

12) $2x + 3y = 6$
$4x + 6y = 12$

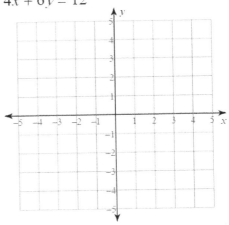

STATISTICS AND PROBABILITY

Statistics concerns data and methods of representing and analyzing data. We will analyze graphs, find statistical measures, and find and interpret probability.

ANALYZING GRAPHS

There are many types of graphs which include pictographs, pie charts, bar charts, line charts, histograms, etc. When working with graphs, make sure to read any titles, legends, keys, axes, and any extra information. Make sure you understand what is being represented and how (as a number, percentage, ranking, etc.).

PRACTICE 42

1. Who saw the fewest movies? How many?

2. Who saw the most movies? How many?

3. How many movies did Alice see?

4. What is the ratio of number of movies Bob saw to the number that Emily saw?

5. Of all the movies represented here, what percentage did Alice watch?

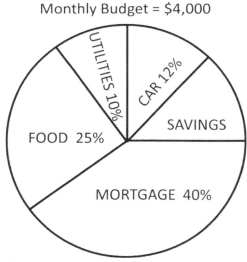

6. How much money is spent on the mortgage?

7. What is the ratio of money spent on food to the money spent on utilities?

8. What percentage goes to savings? How much money is that?

9. If your total monthly budget increases by $200 and you decide to put it all into savings, what percentage is savings now?

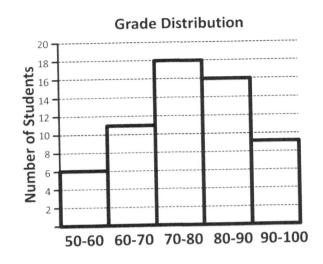

10. If a B is 80-90, then how many students have a B or above?

11. Below a 60 is failing, what is the ratio of passing to failing students?

12. What percentage of students scored between 70 and 80?

DESCRIPTIVE STATS

We can use statistics to quantify characteristics of a data set. Primarily we want to know the center and the spread of the data. The center can be measured using the mean, median, and mode.

Mean, or average: $\bar{x} = \frac{sum\ of\ the\ data\ values}{number\ of\ data\ values}$

Median is the middle number. Put the list in order and find the middle number. If there are an even number of items, then take the average of the two middle numbers.

Mode is the most frequent number. There may be one mode, no mode, or multiple modes.

One way we can measure spread is by the **range** = highest value – lowest value

CONSTRUCTING A BOXPLOT

A boxplot, also called a box and whisker plot, is one way to visual data. First put the data in order and find the median. Then find the first quartile Q_1 by taking the median of all the values below the median. Next, find the third quartile Q_3 by taking the median of all the values above the median. We will also use the minimum (lowest number) and maximum (highest number). We place these 5 points (min, Q_1, med, Q_3, and max) on a number line. Draw a box with sides at Q_1 and Q_3. Draw a vertical line at the median. Draw horizontal line on the left side from Q_1 to the min. Draw another horizontal line on the right side from Q_3 to the max. The interquartile range is another measure of spread.

The **interquartile range (IQR) = $Q_3 - Q_1$**. Let's see all of this in action.

Ex. 1 Create a boxplot for this data set below. Find the mean, median, mode, range, min, Q1, Q3, max, and IQR.

Test Grades: $\{88, 78, 83, 70, 73, 54, 98, 68, 75, 78, 83, 92, 66, 72\}$

First, we put it in order. Find the median. Since there are 14 numbers the median will be between 75 and 78. Look at the numbers left of the median: 54, 66, 68, 70, 72, 73, 75.

$\{54, 66, 68, 70, 72, 73, 75, 78, 78, 83, 83, 88, 92, 98\}$

$Min = 54 \quad Q_1 = 70 \quad med = \dfrac{75+78}{2} = 76.5 \quad Q_3 = 83 \quad Max = 98$

The median of those is 70 which is our Q_1. Look at the numbers right of the median: 78, 78, 83, 83, 88, 92, 98. The median of those is 83 which is our Q_3. The minimum is 54 and the maximum is 98. We draw the boxplot.

Mean = 77
Median = 76.5
Mode = 78, 83
Range = 44
Min = 54
Q1 = 70
Q3 = 83
Max = 98
IQR = 83-70 = 13

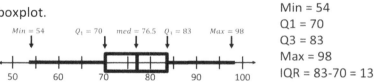

Ex. 2 Create a boxplot for this data set below. Find the mean, median, mode, range, min, Q1, Q3, max, and IQR.

Test Grades: {64, 68, 70, 72, 76, 78, 80, 88, 96} This set is already in order. The median is the middle number which is 76. Q_1 is the median of the numbers to the left of the median (ignore the median) 64, 68, 70, 72. We have to take the average of 68 and 70 to get Q_1 = 69. Q_3 is the median of the numbers to the right of the median 78, 80, 88, 96. We have to take the average of 80 and 88 to get Q_3 = 84. The min is 64 and the max is 96. Draw a number line from 60 to 100. Draw a point at 64, 69, 76, 84, 96 and make the boxplot.

Mean = 76.88
Median = 76
Mode = no mode
Range = 32
Min = 64
Q1 = 69
Q3 = 84
Max = 96
IQR = 84-69 = 13

CALCULATOR HELP: STATISTICS

We can use the graphing calculator to find most of these statistical measures.

Press the **STAT** button and go to `1: Edit` Press **ENTER** This will take you to a screen with lists. We will type our data into L_1. If there's data already there you can either overwrite it or delete the whole list by moving the cursor up to L_1 then press **CLEAR** **ENTER**. Type in the data into L_1 pressing ENTER after each entry. Then press **STAT** This time going over to the CALC menu. It should say `1-Var Stats` at the top. Press **ENTER** a few times (about 4

STATISTICS AND PROBABILITY

times) until you get to a new screen with some symbols and numbers. If you scroll down with the arrow keys, you will get the numbers you need to make a boxplot.

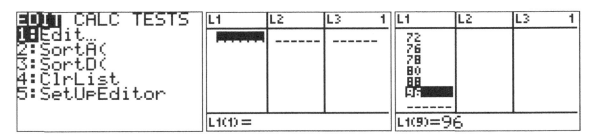

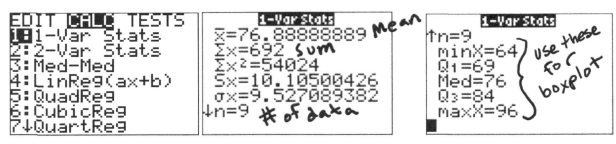

PRACTICE 43

Find the mode, median, mean, lower quartile, upper quartile, and interquartile range for each data set.

1) Hits in a Round of Hacky Sack

6	12	12	10	7	10	19
5	4	15	4	12	5	4
18	3					

2) Shoe Size

6.5	10	7.5	6	5.5	9	9
8	9.5	10	6	7.5	6	9
5.5	7					

3) # Words in Book Titles

| 2 | 2 | 4 | 3 | 4 | 2 | 2 | 3 |
| 4 | 2 | 6 | 2 | 2 | 2 | 3 | |

4) Test Scores

50	51	49	43	49	53	49
47	44	47	53	52	50	36
41						

Draw a box-and-whisker plot for each data set.

5) Minutes to Run 5km

22.1   20.8   22.1   23.2   27.1
36.2   31.5   34.6   36.8

6) Shoe Size

10   7.5   6.5   9   11   8   9.5
8.5   8   7.5   9

7) Hits in a Round of Hacky Sack

4   12   3   3   5   18   6   3
3   1   19

8) Mens Heights (Inches)

71   67   76   67   81   63   68
67   73   69   78

PROBABILITY

Probability is the measure of the likelihood or chance that an event will occur. We will calculate probability as the following ratio: $probability = \frac{number\ of\ desired\ outcomes}{total\ number\ of\ outcomes}$. Therefore, the probability that an event will happen will be a fraction between zero and one. If the probability is zero it is impossible for it to happen; if the probability is one it is certain to happen.

Sometimes it is difficult to count the number of desired outcomes or the total number of outcomes. For example, if we roll a dice and flip a coin, how many ways can this happen? We could list all the possibilities out, but it is much easier to use the **multiplication principle**. The **multiplication principle** states that if there are a ways of doing one thing and b ways of doing a second thing, then there are $a \times b$ ways of doing both. So, with our dice and coin, there are 6 ways to roll a dice and 2 ways to flip a coin, therefore there are 12 ways of rolling a dice and flipping a coin. We can list them out in a **sample space**, the set of all possible outcomes: {1H, 1T, 2H, 2T, 3H, 3T, 4H, 4T, 5H, 5T, 6H, 6T}.

Ex. 1 What is the probability of rolling a number greater than 4 with one dice? There are 2 desired outcomes {5, 6}. There are 6 total outcomes {1, 2, 3, 4, 5, 6}. So $P(roll > 4) = \frac{2}{6} = \frac{1}{3}$.

What if we roll two dice? Again, we can apply the multiplication principle. If there are 6 ways to roll the first dice and 6 ways to roll the second dice, then there should be 36 ways to roll two dice. It's easy to visualize this by making a table of all possibilities.

	1	**2**	**3**	**4**	**5**	**6**
	2	3	4	5	6	7
2	3	4	5	6	7	8
3	4	5	6	7	8	9
4	5	6	7	8	9	10
5	6	7	8	9	10	11
6	7	8	9	10	11	12

In this table, the numbers in bold represent the numbers on the dice and we take the sum of the dice to fill in the table. Which number is most likely? Rolling a 7 is most likely with probability of $\frac{6}{36} = \frac{1}{6}$. What is P(rolling 5 with 2 dice)?

$$P(rolling\ 5\ with\ 2\ dice) = \frac{4}{36} = \frac{1}{9}.$$

Another common scenario with probability is flipping coins. If we flip 1 coin, what is P(Heads)? Clearly, we get $\frac{1}{2}$.

Ex. 2 What is the probability of getting at least 1 head if we flip 2 coins? What is the probability of getting 2 heads with 3 coins? For two coins our sample space is {HH, HT, TH, and TT}. Three of these have at least one head so $P(at\ least\ 1\ H\ with\ 2\ coins) = \frac{3}{4}$. For three coins our sample space is {HHH, HHT, HTH, HTT, THH, THT, TTH, TTT}. 3 of these have exactly 2 heads so $P(2\ H\ with\ 3\ coins) = \frac{3}{8}$.

PRACTICE 44

1. What is probability of rolling an even number with 1 dice?
2. What is the probability of rolling a factor of 12 with 1 dice?
3. What is the probability of rolling a 10 with 1 dice?
4. What is the probability of rolling a number less than 5 with 1 dice?
5. What is probability of rolling an even number with 2 dice?
6. What is the probability of rolling a factor of 12 with 2 dice?
7. What is the probability of rolling a 10 with 2 dice?
8. What is the probability of rolling a number less than 5 with 2 dice?
9. What is the probability of getting exactly 2 tails flipping a coin twice?
10. What is the probability of getting at least 1 tail flipping a coin twice?
11. What is probability of getting two heads in a row flipping a coin twice?
12. What is the probability of getting exactly 2 tails flipping a coin three times?
13. What is the probability of getting at least 1 tail flipping a coin three times?
14. What is probability of getting two heads in a row flipping a coin three times?

SOLUTIONS

PRACTICE 1

1) N, W, Z, Q, R
2) Z, Q, R
3) Q, R
4) I, R
5) W, Z, Q, R
6) N, W, Z, Q, R
7) Q, R
8) I, R
9) N, W, Z, Q, R
10) Z, Q, R
11) Q, R
12) I, R
13) N, W, Z, Q, R
14) N, W, Z, Q, R
15) I, R
16) Q, R
17) Z, Q, R
18) N, W, Z, Q, R
19) W, Z, Q, R
20) W, Z, Q, R

PRACTICE 2

1) 45
2) −9
3) −10
4) 5
5) 110
6) 96
7) 2
8) −4
9) −2
10) −10
11) −28
12) −20
13) 43
14) 11
15) −154
16) −30
17) −11
18) 7
19) 0
20) 21
21) −42
22) 8
23) 18
24) −29

PRACTICE 3

Identify the place value of each underlined digit

1. 26, 3$\underline{4}$1, 798.1467   ten thousands
2. 67, 153.95$\underline{8}$2   thousandths
3. 38.$\underline{0}$19528   tenths
4. 0.294$\underline{8}$3   ten thousandths
5. 1, 5$\underline{8}$2,949, 028.93   ten millions
6. 3.1$\underline{4}$15926535897   hundredths
7. 45.9$\underline{8}$7   hundredths
8. 4, 21$\underline{6}$.3893   ones
9. 8932.$\underline{6}$832   tenths
10. $\underline{2}$, 485, 692.13   thousands
11. 2, 4$\underline{8}$5, 692.13   ten thousands
12. 3.141592653$\underline{5}$897   ten thousandths
13. 513.899$\underline{7}$31   thousandths
14. 513.8$\underline{9}$9731   hundredths
15. 51$\underline{3}$.899731   ones
16. 26, 341, 79$\underline{8}$.1467   tens
17. 26, 341, 79$\underline{8}$.1467   ones
18. $\underline{3}$.1415926535897   ones
19. 3.1415$\underline{9}$26535897   hundred thousandths
20. $\underline{2}$6, 341, 798.1467   millions

PRACTICE 4

Round to indicated place value

1. 96, 400, 000
2. 64.25
3. 59, 300
4. 13, 590.0025
5. 1.9
6. 45, 938
7. 54, 024.249
8. 94, 430
9. 3.14
10. 3.142
11. 3, 900
12. 3, 900
13. 4, 000
14. 3.1

SOLUTIONS

15. 3
16. 54, 000
17. 96, 400, 000

18. 96, 399, 530
19. 799
20. 800

PRACTICE 5

1) 5.9
2) 6.084
3) 8.52
4) 0.583
5) 8.33
6) 12.636
7) 1.9
8) 0.4
9) 1.68
10) 3.432
11) 1.615
12) 0.414
13) 7.588
14) 3.38
15) 1.85
16) 47
17) 30.08
18) 53.01
19) 15.2
20) 15.5
21) 3.5
22) 0.5
23) 9.75
24) 5

PRACTICE 6

1. 40, 24 LCM = 120 GCF = 8
2. 21, 44 LCM = 924 GCF = 1
3. 28, 20 LCM = 140 GCF = 4
4. 60, 35 LCM = 420 GCF = 5
5. 48, 30 LCM = 240 GCF = 6
6. 45, 72 LCM = 360 GCF = 9
7. 56, 42 LCM = 168 GCF = 14
8. 45, 10 LCM = 90 GCF = 5
9. 32, 36 LCM = 288 GCF = 4
10. 21, 42 LCM = 42 GCF = 21
11. 56, 48 LCM = 336 GCF = 8
12. 16, 36 LCM = 144 GCF = 4
13. 48, 16 LCM = 48 GCF = 16
14. 20, 46 LCM = 460 GCF = 2
15. 40, 24, 16 LCM = 240 GCF = 8
16. 18, 12, 30 LCM = 180 GCF = 6
17. 45, 75, 30 LCM = 450 GCF = 15
18. 7, 8, 9 LCM = 504 GCF = 1
19. 56, 20, 40 LCM = 280 GCF = 4
20. 27, 16, 32 LCM = 864 GCF = 1

PRACTICE 7

1) 2
2) $\frac{7}{6}$
3) $\frac{51}{10}$
4) $\frac{20}{3}$
5) $\frac{9}{4}$
6) $\frac{107}{20}$
7) $\frac{4}{7}$
8) $\frac{1}{12}$
9) $\frac{1}{4}$
10) $\frac{11}{12}$
11) $\frac{11}{24}$
12) $\frac{5}{8}$
13) $\frac{27}{25}$
14) $\frac{6}{5}$
15) $\frac{17}{3}$
16) $\frac{1}{6}$
17) $\frac{21}{8}$
18) $\frac{28}{3}$
19) $\frac{10}{9}$
20) $\frac{10}{7}$
21) $\frac{7}{3}$
22) $\frac{21}{100}$
23) $\frac{12}{7}$
24) $\frac{14}{27}$

SOLUTIONS

PRACTICE 8

1) 11
2) −24
3) −9
4) 8
5) 13
6) −6
7) 3
8) 32
9) 4
10) 37

11) 6
12) 31.83
13) −3.7
14) −0.46
15) 0.12
16) 8.16
17) 14.28
18) 38.85
19) −8.25
20) 18.94
21) $-\dfrac{13}{4}$
22) $\dfrac{5}{6}$
23) $\dfrac{44}{15}$
24) $\dfrac{9}{4}$
25) $\dfrac{25}{3}$
26) $\dfrac{3}{10}$
27) $-\dfrac{5}{2}$
28) $\dfrac{5}{2}$
29) $\dfrac{29}{12}$
30) $-\dfrac{1}{8}$

PRACTICE 9

1) −12
2) −16
3) 21
4) −1
5) 4
6) 16
7) 1
8) 6
9) 36
10) −1

11) 5.72
12) −8.68
13) 0.72
14) 46.35
15) $\dfrac{5}{6}$
16) $\dfrac{43}{10}$
17) $\dfrac{71}{30}$
18) $\dfrac{1}{4}$
19) $-\dfrac{31}{20}$
20) $-\dfrac{6}{5}$

PRACTICE 10

1. 0.8
2. $.\overline{6}$
3. 0.875
4. 2.5
5. 1.75
6. 2
7. 0.4
8. 0.6
9. 0.02
10. 0.003
11. 0.9
12. 1.4
13. $\dfrac{1}{3}$
14. $\dfrac{7}{4} = 1\dfrac{3}{4}$
15. $\dfrac{3}{5}$
16. $\dfrac{1}{4}$
17. $\dfrac{1}{50}$
18. $\dfrac{12}{5} = 2\dfrac{2}{5}$
19. $\dfrac{1}{5}$
20. $\dfrac{3}{4}$
21. $\dfrac{1}{50}$
22. 3
23. $\dfrac{3}{5}$
24. $\dfrac{9}{5} = 1\dfrac{4}{5}$
25. 74%
26. 205%
27. 30%
28. 500%

SOLUTIONS

29. 35%
30. 130%
31. 50%

32. 120%
33. 25%
34. 37.5%

35. 40%
36. 150%

PRACTICE 11

1) true
2) false
3) false
4) true
5) {15}
6) {6}
7) $\left\{\dfrac{14}{3}\right\}$
8) $\left\{\dfrac{9}{5}\right\}$
9) $\left\{\dfrac{15}{2}\right\}$
10) {10}
11) $\left\{\dfrac{24}{7}\right\}$
12) $\left\{\dfrac{12}{5}\right\}$
13) $\left\{\dfrac{45}{2}\right\}$
14) $\left\{\dfrac{5}{2}\right\}$
15) {15}
16) $\left\{\dfrac{32}{5}\right\}$
17) $\left\{\dfrac{63}{2}\right\}$
18) {21}
19) $\left\{\dfrac{14}{3}\right\}$
20) $\left\{\dfrac{5}{3}\right\}$

PRACTICE 12

1. .02 L
2. 400 cg
3. 4 cg
4. 40 g
5. 200 mL
6. 4000 cm
7. .0042 g
8. 0.0024 kg
9. 562 cm
10. 1,860 mm
11. 7.18 m
12. 88,140 m
13. 54,410 mL
14. 48.71 L
15. 8,550 mg
16. 69.5 g
17. 7,880,000 mg
18. .002 km
19. 2040 mL
20. 3200 mL

PRACTICE 13

1) 60.8%
2) 122.6%
3) 74.2
4) 20.9
5) 40.3
6) 27.1
7) 9.8%
8) 3.9
9) 41.1
10) 97.8%
11) 5.7 hours
12) 25.5%
13) 82.9 m
14) 126 miles
15) 27.9%
16) 4.2 miles
17) 13.8%
18) 1043.8 m
19) 397.3 miles
20) 385.7%

SOLUTIONS

PRACTICE 14

1) 8.2% decrease
2) 62.5% increase
3) 111.1% increase
4) 43.8% decrease
5) 68.7% decrease
6) 2.4% decrease
7) 26.6% increase
8) 250% increase
9) 8.6% decrease
10) 314.3% increase
11) 66.7% decrease
12) 60.9% decrease
13) 98.6% decrease
14) 125.8% increase
15) 37.5% decrease
16) 124% increase
17) 171.4% increase
18) 60.3% decrease
19) 39.6% decrease
20) 40% decrease

PRACTICE 15

1. $71.99
2. $44.93
3. $24.43
4. $35.99
5. $166.74
6. $21.21
7. $34.86
8. $76.43
9. $62.39
10. $26.70
11. $260.71
12. $68.92
13. $83.99
14. $630.65
15. $87.92
16. $91.40

PRACTICE 16

1. maturity value: $14,375 monthly payment: $399.31
2. maturity value: $20,732 monthly payment: $246.81
3. maturity value: $3030 monthly payment: $67.33
4. maturity value: $1955 monthly payment: $65.17
5. maturity value: $787,836 monthly payment: $2188.43
6. maturity value: $21,543 monthly payment: $359.05
7. maturity value: $2455.25 monthly payment: $136.40
8. maturity value: $16972.5 monthly payment: $565.75
9. maturity value: $1690.31 monthly payment: $80.49
10. maturity value: $30420 monthly payment: $507

PRACTICE 17

1. $5.08
2. 4 packages
3. 6 bars
4. $2.79
5. 14 euros
6. $15.67
7. $2.30
8. $5.58
9. 20
10. 15
11. 12.5 inches
12. 3 cm

Page | 174

SOLUTIONS

PRACTICE 18

1. 34
2. 70
3. 155
4. 148
5. Acute
6. Straight
7. Obtuse
8. Right
9. Obtuse
10. Right
11. Straight
12. Acute
13. Vertical, 35
14. Adjacent, 36
15. Supplementary/Linear pair, 142
16. Complementary, 57
17. Vertical, 34
18. Supplementary/Linear pair, 132
19. Adjacent, 39
20. Complementary, 35

PRACTICE 19

1. Corr., 133
2. Vert., 91
3. Alt. Int., 50
4. S-S Int., 122
5. Corr., 93
6. Alt. Int., 108
7. Alt. Ext., 125
8. S-S Int., 84
9. Alt. Int., 60
10. Alt. Ext., 128

PRACTICE 20

1)  2) Not possible 3)  4)

5) Not possible 6)  7) obtuse scalene

8) right isosceles 9) acute isosceles 10) right scalene 11) obtuse isosceles
12) equilateral
13) acute isosceles 14) right scalene 15) right isosceles 16) equilateral
17) acute scalene 18) obtuse isosceles

Page | 175

SOLUTIONS

PRACTICE 21

1) 122°
2) 30°
3) 75°
4) 60°
5) 110°
6) 61°
7) 91°
8) 140°
9) 125°
10) 95°
11) 130°
12) 107°

PRACTICE 22

1) quadrilateral
2) trapezoid
3) kite
4) parallelogram
5) isosceles trapezoid
6) rhombus
7) kite
8) rectangle
9) isosceles trapezoid
10) square

PRACTICE 23

1) pentagon
2) quadrilateral
3) hexagon
4) decagon
5) heptagon
6) nonagon
7) octagon
8) hexagon
9) nonagon
10) pentagon
11) decagon
12) quadrilateral

PRACTICE 24

1) pentagonal pyramid
2) triangular pyramid
3) square pyramid
4) sphere
5) square prism
6) cone
7) hexagonal pyramid
8) rectangular prism
9) rectangular pyramid
10) pentagonal prism
11) square prism
12) hexagonal prism
13) triangular pyramid
14) cylinder
15) rectangular prism
16) trapezoidal prism
17) sphere
18) hexagonal pyramid
19) triangular prism
20) rectangular pyramid

PRACTICE 25

1)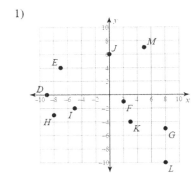

2) $N(4, -5)$ $M(1, 1)$ $L(8, 4)$
   $K(-2, -1)$ $J(10, -2)$ $I(-5, 3)$
   $H(6, 3)$ $G(-5, 10)$ $F(-9, -4)$
   $E(4, -6)$

SOLUTIONS

PRACTICE 26

1) $(-4, -2)$
2) $(1, 0)$
3) $\left(2, \dfrac{1}{2}\right)$
4) $\left(-1\dfrac{1}{2}, -2\right)$
5) $(-3, 7)$
6) $(3, 5)$
7) $\left(-11, \dfrac{1}{2}\right)$
8) $\left(-4\dfrac{1}{2}, -11\right)$
9) $(5.85, 1.3)$
10) $(0.25, 3.9)$
11) $\left(1\dfrac{7}{8}, 1\dfrac{8}{9}\right)$
12) $\left(3\dfrac{1}{7}, 2\dfrac{2}{9}\right)$

PRACTICE 27

1) 15 in
2) 20 yd
3) 14.3 ft
4) 11.1 cm
5) 10 in
6) 12.9 in
7) 9.6 m
8) 10.8 km
9) 5 cm
10) 15 m
11) 10 cm
12) 9 mi
13) 6.9 m
14) 7.3 km
15) 10.3 mi
16) 3.7 in

PRACTICE 28

1) 8.5
2) 7.3
3) 5.4
4) 9.2
5) 3.6
6) 8.2
7) 11.2
8) 4.1
9) 14.8
10) 6.8
11) 14.9
12) 3.8
13) 6.4
14) 8.2
15) 10
16) 13.9
17) 14.9
18) 15.3
19) 13.6
20) 8.1

PRACTICE 29

1. P = 22.6 m; A = 31.5 m²
2. P = 25.2 m; A = 28.13 cm²
3. P = 72.986 m; A = 109 m²
4. P = 51 ft; A = 56.28 ft²
5. P = 28.4 m; A = 34.775 m²
6. P = 25 km; A = 28.8 km²
7. C = 50.27 yd; A = 201.1 yd²
8. C = 12.57 m; A = 12.6 m²
9. P = 25.1 m; A = 28.9 m²
10. P = 15 m; A = 13.16 m²
11. P = 41.8 m; A = 93.5 m²
12. P = 44.448 m; A = 76.5 m²
13. P = 22.4 ft; A = 28.8 ft²
14. P = 28 cm; A = 35.75 cm²
15. C = 18.85 m; A = 28.3 cm²
16. C = 62.83 m; A = 314.2 m²
17. P = 22.6 m; A = 24.4 m²
18. P = 27 m; A = 31.35 m²
19. P = 11 m; A = 5.5 m²
20. P = 7 mi; A = 2.4 mi²

SOLUTIONS

PRACTICE 30

1. P = 65.7 m; A = 239.3 m²

2. P = 85.7 m; A = 260.7 m²

3. P = 36 ft; A = 72 ft²

PRACTICE 31

1) ASA
2) SAS
3) ASA
4) Not enough information
5) SAS
6) AAS
7) SSS
8) ASA
9) Not enough information
10) SSS
11) ASA
12) Not enough information
13) Not enough information
14) SSS
15) SAS
16) SAS
17) Not enough information
18) AAS
19) Not enough information
20) SAS

PRACTICE 32

1) similar; SSS similarity
2) not similar
3) similar; SAS similarity
4) similar; SSS similarity
5) similar; AA similarity
6) similar; SSS similarity
7) similar; AA similarity
8) not similar
9) 24
10) 13
11) 45
12) 15

PRACTICE 33

1) 75.4 km³
2) 2211.68 mi³
3) 405 ft³
4) 523.6 yd³
5) 402.12 mi³
6) 2094.4 km³
7) 280 m³
8) 113.1 km³
9) 3619.11 ft³
10) 960 cm³
11) 4188.79 m³
12) 923.63 ft³

PRACTICE 34

1) $11b - 3$
2) $2m + 6$
3) $6n + 7$
4) -6
5) $-6p - 42$
6) $3a + 30$
7) $15p - 3$
8) $-14b + 28$
9) $2m + 16$
10) $1 - 20x$
11) $-12x - 31$
12) $-x + 45$
13) $-2p - 3$
14) $9 + 18n$
15) $13x + 3$
16) $9x + 33$
17) $20m - 5$
18) $-58k + 80$
19) $-18 - 12x$
20) $-2 + 49n$

SOLUTIONS

PRACTICE 35

1) 8
2) −2
3) 6
4) 20
5) −31
6) −26
7) 56
8) −30

PRACTICE 36

1) {10}
2) {−14}
3) {17}
4) {72}
5) {−19}
6) {2}
7) {2}
8) {−6}
9) {−14}
10) {−12}
11) {20}
12) $\left\{\dfrac{3}{5}\right\}$
13) {6}
14) $\left\{-\dfrac{7}{2}\right\}$

PRACTICE 37

1) {−4}
2) {1}
3) {7}
4) {−7}
5) {−5}
6) {2}
7) {6}
8) {−2}
9) $\left\{\dfrac{5}{2}\right\}$
10) $\left\{\dfrac{13}{4}\right\}$
11) $\left\{\dfrac{4}{3}\right\}$
12) $\left\{-\dfrac{5}{4}\right\}$
13) {1}
14) {−2}
15) {−7}
16) {−1}
17) {4}
18) {5}
19) {−6}
20) {7}
21) $\left\{-\dfrac{3}{2}\right\}$
22) $\left\{-\dfrac{5}{4}\right\}$

PRACTICE 38

1) $a < -5$:
2) $n \leq -16$:
3) $k > -9$:
4) $a \leq -16$:
5) $m < -1$:
6) $m < 4$:
7) $b > -1$:
8) $a \leq 6$:
9) $v > -7$:
10) $n > 0$:
11) $n \geq 0$:
12) $x > 0$:
13) $x > -\dfrac{1}{2}$:
14) $b > \dfrac{2}{3}$:

SOLUTIONS

15) $x < -2$:
17) $k > 1$:
19) $n > 2$:
21) $x > 2$:
23) $v < -\dfrac{5}{3}$:

16) $n \geq -2$:
18) $p \leq -2$:
20) $n < -8$:
22) $n > 0$:
24) $x \geq \dfrac{1}{3}$:

PRACTICE 39

1) -1
2) Undefined
3) $\dfrac{1}{4}$
4) 1

5) 0
6) -2
7) $\dfrac{8}{5}$
8) $-\dfrac{3}{4}$

9) Undefined
10) $\dfrac{5}{2}$
11) 1
12) $\dfrac{1}{2}$

13) $-\dfrac{7}{3}$
14) -1
15) $\dfrac{9}{2}$
16) 0

PRACTICE 40

1)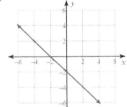
2)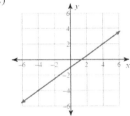
3)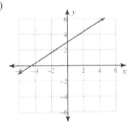
4)
5)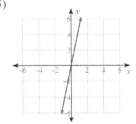
6)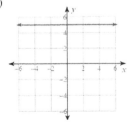

SOLUTIONS

7) 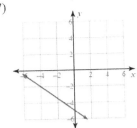 8) 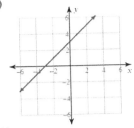 9)

10) 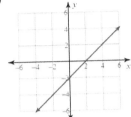 11) 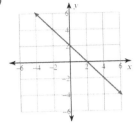 12)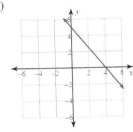

13) 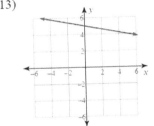 14)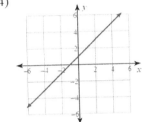

PRACTICE 41

1) $(4, -2)$
2) $(2, -1)$
3) $(2, -3)$
4) $(4, 3)$
5) $(2, 1)$
6) $(2, 2)$
7) $(2, -4)$
8) $(-2, -3)$
9) $(-1, 1)$
10) No solution
11) $(4, -3)$
12) Infinite Solutions

PRACTICE 42

1. Emily 12
2. John 32
3. 22 (the half box counts as 2)
4. 7:3
5. 5.5/23.5 = 23.4%
6. .4(4000)=1600
7. 25/10=5/2
8. 13%; $520

SOLUTIONS

9. $\frac{520+200}{4000+200} = \frac{720}{4200} = .1714 = 17.1\%$

10. 25

11. 54/6 = 9 to 1, 9:1, or $\frac{9}{1}$

12. 18/60 = 30%

PRACTICE 43

1) Mode = 4 and 12, Median = 8.5, Mean = 9.13, $Q_1 = 4.5$, $Q_3 = 12$ and IQR = 7.5

2) Mode = 6 and 9, Median = 7.5, Mean = 7.63, $Q_1 = 6$, $Q_3 = 9$ and IQR = 3

3) Mode = 2, Median = 2, Mean = 2.87, $Q_1 = 2$, $Q_3 = 4$ and IQR = 2

4) Mode = 49, Median = 49, Mean = 47.6, $Q_1 = 44$, $Q_3 = 51$ and IQR = 7

5)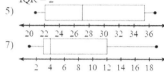

6)

7)

8)

PRACTICE 44

1. 1/2
2. 5/6
3. 0
4. 2/3
5. 1/2
6. 1/3
7. 1/12
8. 1/6
9. 1/4
10. 3/4
11. 1/4
12. 3/8
13. 7/8
14. 3/8

APPENDIX I: TABLES
Table 1: Calculator Quick Reference

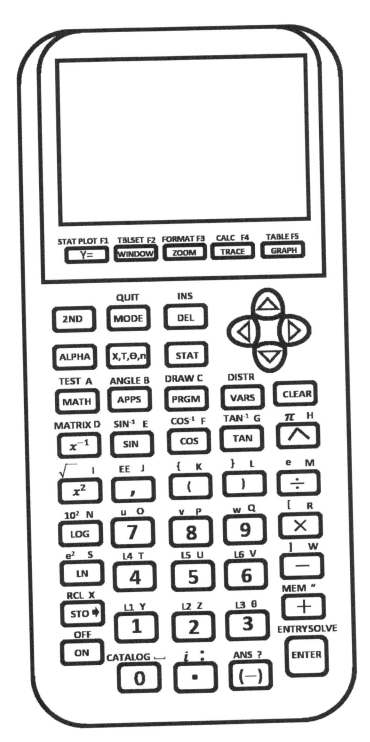

APPENDIX I: TABLES

TEST A
[MATH]

```
MATH NUM CPX PRB
1:▶Frac
2:▶Dec
3:³
4:³√(
5:ˣ√
6:fMin(
7↓fMax(
```

STAT

```
EDIT CALC TESTS
1:Edit…
2:SortA(
3:SortD(
4:ClrList
5:SetUpEditor
```

```
MATH NUM CPX PRB
8↑lcm(
9:gcd(
0:remainder(
A:▶n/d◀▶Un/d
B:▶F◀▶D
C:Un/d
D:n/d
```

```
EDIT CALC TESTS
1:1-Var Stats
2:2-Var Stats
3:Med-Med
4:LinReg(ax+b)
5:QuadReg
6:CubicReg
7↓QuartReg
```

STAT PLOT F1
[Y=]

TABLE F5
[GRAPH]

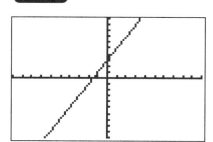

TABLE F5
[2ND] [GRAPH]

CALC F4
[2ND] [TRACE]

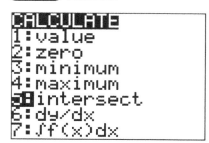

```
CALCULATE
1:value
2:zero
3:minimum
4:maximum
5:intersect
6:dy/dx
7:∫f(x)dx
```

Table 2: Venn Diagram of Real Numbers

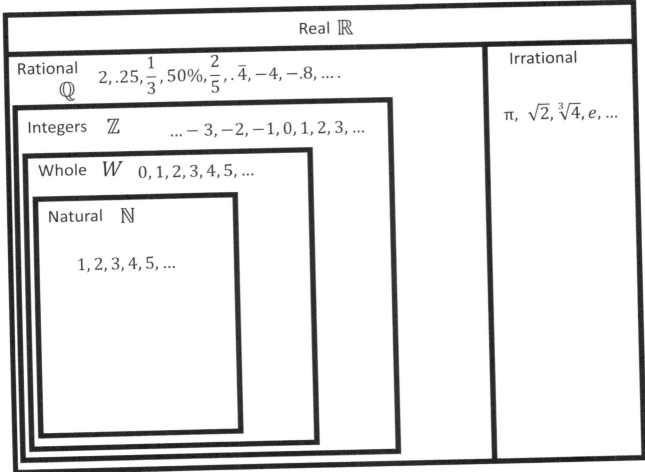

APPENDIX I: TABLES

Table 3: Decimal System

Base 10 (decimal) system

Examples	million	hundred thousand	ten thousand	thousand	hundred	ten	unit	decimal point	tenth	hundredth	thousandth	ten thousandth	hundred thousandth
1 million	1	0	0	0	0	0	0	•					
400 thousands		4	0	0	0	0	0	•					
3 thousand 2 hundred				3	2	0	0	•					
6 hundred seventeen					6	1	7	•					
Twenty						2	0	•					
5 tenths							0	•	5				
8 hundredths							0	•	0	8			
73 hundred thousandths							0	•	0	0	0	7	3

Table 4: Large Number Names

Million	1,000,000	10^6
Billion	1,000,000,000	10^9
Trillion	1,000,000,000,000	10^{12}
Quadrillion	1,000,000,000,000,000	10^{15}
Quintillion	1,000,000,000,000,000,000	10^{18}
Sextillion	1,000,000,000,000,000,000,000	10^{21}
Septillion	1,000,000,000,000,000,000,000,000	10^{24}
Octillion	1,000,000,000,000,000,000,000,000,000	10^{27}
Nonillion	1,000,000,000,000,000,000,000,000,000,000	10^{30}
Decillion	1,000,000,000,000,000,000,000,000,000,000,000	10^{33}

APPENDIX I: TABLES

Table 5: Divisibility Rules

	A number is divisible by...	Divisible	Not Divisible
2	The last digit is even (0,2,4,6,8)	201**8**	2019
3	The sum of the digits is divisible by 3	2019 2+0+1+9=12	2020
4	The last 2 digits are divisible by 4	20**20**	2021
5	The last digit is 0 or 5	202**0**	2019
6	Is even and is divisible by 3 (follows 2 rule and 3 rule)	201**6** 2+0+1+6=9	2019
7	Double the last digit and subtract it from a number made by the other digits. The result must be divisible by 7.	2016 201-12=189 18-18=0	2017
8	The last three digits are divisible by 8	2**024**	2020
9	The sum of the digits is divisible by 9	2025 2+0+2+5=9	2019
10	The number ends in 0	202**0**	2021
11	Add and subtract digits in an alternating pattern (add digit, subtract next digit, add next digit, etc.). Then check if that answer is divisible by 11.	2035 2-0+3-5=0	2037
12	The number is divisible by both 3 and 4 (follows the 3 rule and 4 rule)	20**28** 2+0+2+8=12	2037 Divisible by 3 but not 4
13	Truncate the last digit, multiply it by 4 and add it to the rest of the number. The result is divisible by 13 if and only if the original number was divisible by 13. This process can be repeated for large numbers.	299 29+(9x4) = 29+36 = 65 6+(5x4) = 26 26 is divisible	204 20+(4x4) = 36 204 not divisible
14	Follows the rules for 2 and 7	2016	637
15	Follows the rules for 3 and 5	2115	2015
16	A number is divisible by 16 if the thousands digit is even, and the last three digits form a number that is divisible by 16.	2016	3016
17	Subtract 5 times the last digit from the rest of the number, if the result is **divisible** by **17** then the number is also **divisible** by **17**.	221 22-(5x1) = 17	317 31-(5x7) = -4
18	Follows the rules for 9 and 2	5202	1348
19	Multiply the last digit by 2 and to the rest of the number, if the result is divisible by 19	3059 305+(9x2) = 323 32+(3x2) = 38 3+(2x8) = 19	3072 307+(2x2) =311 31+(2x1) =33 3+(2x3) =9

Table 6: Table of Prime Numbers

1	2	3	4	5	6	7	8	9	10
11	12	13	14	15	16	17	18	19	20
21	22	23	24	25	26	27	28	29	30
31	32	33	34	35	36	37	38	39	40
41	42	43	44	45	46	47	48	49	50
51	52	53	54	55	56	57	58	59	60
61	62	63	64	65	66	67	68	69	70
71	72	73	74	75	76	77	78	79	80
81	82	83	84	85	86	87	88	89	90
91	92	93	94	95	96	97	98	99	100

The white spaces are prime numbers.

APPENDIX I: TABLES

Table 7: Common Percent, Decimal, Fraction Conversions

\multicolumn{3}{c}{Table of Common Conversions}		
Percent	Decimal	Fraction
1%	.01	$\frac{1}{100}$
5%	.05	$\frac{1}{20}$
8.3%	.083	$\frac{1}{12}$
10%	.1	$\frac{1}{10}$
12.5%	.125	$\frac{1}{8}$
$16\frac{2}{3}$%	$.1\overline{6}$	$\frac{1}{6}$
20%	.2	$\frac{1}{5}$
25%	.25	$\frac{1}{4}$
$33\frac{1}{3}$%	$.\overline{3}$	$\frac{1}{3}$
50%	.5	$\frac{1}{2}$
$66\frac{2}{3}$%	$.\overline{6}$	$\frac{2}{3}$
75%	.75	$\frac{3}{4}$
100%	1	1
125%	1.25	$1\frac{1}{4}$
150%	1.5	$1\frac{1}{2}$

APPENDIX I: TABLES

Table 8: Metric System

Prefix	Symbol	Multiplier	
yotta-	Y	10^{24}	1,000,000,000,000,000,000,000,000
zetta-	Z	10^{21}	1,000,000,000,000,000,000,000
exa-	E	10^{18}	1,000,000,000,000,000,000
peta-	P	10^{15}	1,000,000,000,000,000
tera-	T	10^{12}	1,000,000,000,000
giga-	G	10^{9}	1,000,000,000
mega-	M	10^{6}	1,000,000
kilo-	k	10^{3}	1,000
hecto-	h	10^{2}	100
deka-	da	10	10
deci-	d	10^{-1}	0.1
centi-	c	10^{-2}	0.01
milli-	m	10^{-3}	0.001
micro-	μ	10^{-6}	0.000001
nano-	n	10^{-9}	0.000000001
pico-	p	10^{-12}	0.000000000001
femto-	f	10^{-15}	0.000000000000001
atto-	a	10^{-18}	0.000000000000000001
zepto-	z	10^{-21}	0.000000000000000000001
yocto-	y	10^{-24}	0.000000000000000000000001

KING	HENRY	DIED	BY	DRINKING	CHOCOLATE	MILK
KILO-	HECTO-	DEKA-	BASE m, L, g	DECI-	CENT-	MILLI-

APPENDIX I: TABLES

Table 9: Geometry Terms

Term	Definition	Picture
Point	A location in space, having 0 dimensions, no length, no width, no height. Represented by a dot and labeled by a capital letter	• A
Line	Has 1 dimension, length but no width or height. A line extends forever in two directions. Represented by a line with arrows at both ends. Labeled by a script letter or by two points on the line.	ℓ , $\overleftrightarrow{AB}$
Line segment	Part of a line with two endpoints, a beginning and an end.	$\overline{AB}$
Ray	Part of a line with one endpoint and goes on forever in the opposite direction.	$\overrightarrow{AB}$
Plane	Flat 2-dimensional object with length and width like an infinite chalkboard.	P
Polygon	A flat 2-dimensional shape formed by lines connected together. Common polygons include triangles, stars, rectangles, hexagons, etc.	
Solid	A 3-dimensional object having length, width, and height. Common solids include cubes, cones, cylinders, spheres, etc.	
Parallel lines	Lines that never cross. They are always the same distance apart. They also have the same **slope** (discussed later). Parallel lines are marked by a small triangle placed on the line.	$\ell \parallel m$
Intersecting lines	In Euclidean plane geometry, lines are either parallel or they intersect in one point.	

Table 10: Angles

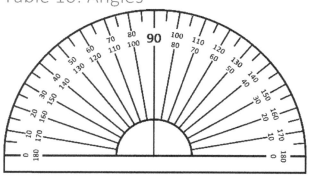

TERM	DEFINITION	PICTURE EXAMPLE
Right angle	90° angle marked by a square at the vertex	
Perpendicular lines	Lines that form a right angle. Symbol for perpendicular is ⊥	$\ell \perp m$
Acute angle	less than 90 degrees	
Obtuse angle	more than 90 degrees	
Straight angle	Equal to 180 degrees. A straight angle forms a straight line.	
Complementary angles	Two angles that add up to 90 degrees	
Supplementary angles Linear Pair	Two angles that add up to 180 degrees	
Vertical angles	Two angles formed by intersecting lines. Vertical angles are congruent (same)	
Adjacent angles	Angles that are "next" to each other. Adjacent angles have the same endpoint and have a common side.	

Table 11: Triangles

	CLASSIFY TRIANGLES BY ANGLES		
TERM	Right	Acute	Obtuse
DEFINITION	Triangle with 1 right angle	Triangle with all acute angles	Triangle with 1 obtuse angle
PICTURE			

	CLASSIFY TRIANGLES BY SIDES		
TERM	Equilateral	Isosceles	Scalene
DEFINITION	Triangle with 3 equal sides (it will also have 3 equal angles, equiangular)	Triangle with 2 equal sides (base angles will also be equal)	Triangle with no equal sides
PICTURE	We put marks on sides to show $\cong$		

Table 12: Quadrilaterals

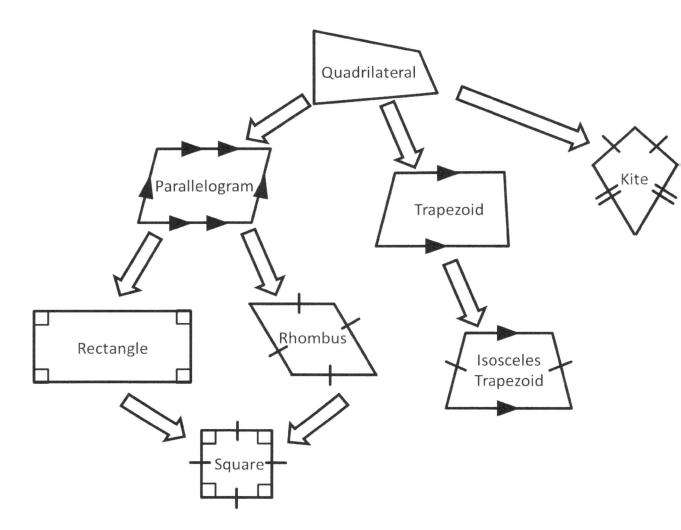

APPENDIX I: TABLES

Table 13: Polygons

Number of Sides	Name	Picture
3	Triangle	
4	Quadrilateral	
5	Pentagon	
6	Hexagon	
7	Heptagon	
8	Octagon	
9	Nonagon	
10	Decagon	
12	Dodecagon	

Table 14: Solids

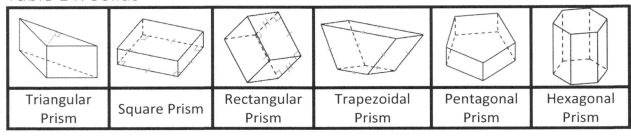

| Triangular Prism | Square Prism | Rectangular Prism | Trapezoidal Prism | Pentagonal Prism | Hexagonal Prism |

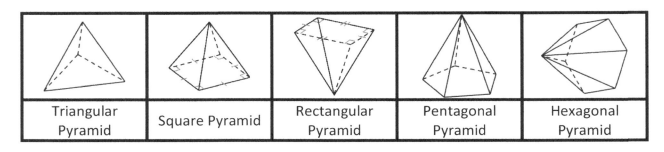

| Triangular Pyramid | Square Pyramid | Rectangular Pyramid | Pentagonal Pyramid | Hexagonal Pyramid |

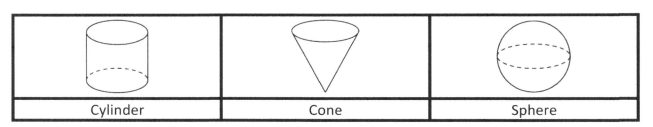

| Cylinder | Cone | Sphere |

Volume

Rectangular Prism	Cube	Sphere	Cylinder	Cone
$V = lwh$	$V = s^3$	$V = \dfrac{4}{3}\pi r^3$	$V = \pi r^2 h$	$V = \dfrac{1}{3}\pi r^2 h$

Table 15: Algebraic Properties

NAME	DEFINITION	EXAMPLES
Commutative	You can change the order in addition and multiplication	$2 + 3 = 3 + 2$ $2 + x = x + 2$ $2 \cdot 3 = 3 \cdot 2$ $x \cdot 3 = 3x$
Associative	You can change the grouping (parentheses) in addition and multiplication	$(2 + 3) + 4 = 2 + (3 + 4)$ $(x + 3) + 4 = x + (3 + 4)$ $(2 \cdot 3)4 = 2(3 \cdot 4)$ $2(3x) = (2 \cdot 3)x$
Distributive	You can multiply a sum by multiplying each addend separately and then add the products.	$2(3 + 4) = 2 \cdot 3 + 2 \cdot 4$ $2(x + 4) = 2x + 2 \cdot 4$
Identity	When you add zero it doesn't change. When you multiply by one it doesn't change	$3 + 0 = 3$ $y + 0 = y$ $1 \cdot 5 = 5$ $1n = n$
Inverse	When you add the opposite, you get zero. When you multiply by the reciprocal you get 1.	$3 + -3 = 0$ $z - z = 0$ $\frac{1}{5} \cdot 5 = 1$ $\frac{1}{n} n = 1$

APPENDIX II: FORMULAS

NAME	FORMULA	PAGE NUMBER
Percent of change	$percent\ change = \dfrac{difference}{original}$	53
Interest	$A = Prt$	59
Maturity Value	$maturity\ value = interest + principal$	59
Monthly Payment	$monthly\ payment = \dfrac{maturity\ value}{number\ of\ months}$	59
Midpoint Formula	$Midpoint\ M = \left(\dfrac{x_1 + x_2}{2}, \dfrac{y_1 + y_2}{2}\right)$	95
Pythagorean Theorem	$a^2 + b^2 = c^2$	97
Distance Formula	$Distance = \sqrt{(x_1 - x_2)^2 + (y_1 - y_2)^2}$	101
Perimeter of Rectangle	$P = 2l + 2w$	104
Perimeter of Square	$P = 4s$	104
Perimeter of Triangle	$P = a + b + c$ add up all sides	104
Circumference of a Circle	$C = 2\pi r$	104
Area of Rectangle	$A = lw$	104

APPENDIX II: FORMULAS

Area of Square	$A = s^2$	104
Area of Parallelogram	$A = bh$	105
Area of Triangle	$A = \frac{1}{2}bh$	105
Area of a Circle	$A = \pi r^2$	106
Volume of Rectangular Prism	$V = lwh$	123
Volume of Cylinder	$V = \pi r^2 h$	123
Volume of Cone	$V = \frac{1}{3}\pi r^2 h$	123
Volume of Sphere	$V = \frac{4}{3}\pi r^3$	123
Slope Definition	$m = \frac{rise}{run}$	143
Slope Formula	$m = \frac{y_2 - y_1}{x_2 - x_1}$	146
Slope-intercept Formula for Line	$y = mx + b$	148
Mean	$\bar{x} = \frac{sum\ of\ data\ values}{number\ of\ data\ values}$	163
Median	$Med = middle\ number$	163
Mode	$Med = Most\ frequent\ number$	163

APPENDIX II: FORMULAS

Range	$Range = Max - Min$	163
Interquartile Range (IQR)	$IQR = Q_3 - Q_1$	164
Probability	$Probability = \dfrac{number\ of\ desired\ outcomes}{total\ outcomes}$	167

REFERENCES

Alfed, P. (1996, August 16). *Why are there infinitely many prime numbers?* Retrieved from University of Utah: Understanding Mathematics: https://www.math.utah.edu/~pa/math/q2.html

Bailey, D. H., Borwein, J. M., Borwein, P., & Plouffe, S. (1997). The quest for PI. *The Mathematical Intelligencer, 19*(1), 50-56. Retrieved 5 18, 2019, from http://crd-legacy.lbl.gov/~dhbailey/dhbpapers/pi-quest.pdf

Caldwell, C. K. (1996). *Primes -- A brief introduction*. Retrieved from The Prime Pages: https://primes.utm.edu/primes/background/introduction.php

Matson, J. (n.d.). *The Origin of Zero*. Retrieved 5 18, 2019, from Springer Nature: http://www.scientificamerican.com/article/history-of-zero/

Maxfield, C. M. (2009). *Alternative Number Systems*. Retrieved 5 18, 2019, from https://sciencedirect.com/science/article/pii/b9781856175074000073

MindYourDecisions. (2016, August 31). *What is 6÷2(1+2) = ? The Correct Answer Explained*. Retrieved May 18, 2019, from YouTube: https://www.youtube.com/watch?v=URcUvFIUIhQ

Norton, J. D. (2017, February 14). *Euclidean Geometry*. Retrieved May 29, 2019, from Einstein for Everyone: https://www.pitt.edu/~jdnorton/teaching/HPS_0410/chapters/non_Euclid_Euclid/index.html

Number Systems. (n.d.). Retrieved 5 18, 2019, from http://www.math.wichita.edu/history/topics/num-sys.html#greek

NumberGenerator.org. (n.d.). *Random 4 Digit Number Generator*. Retrieved from https://numbergenerator.org/random-4-digit-number-generator

Pierce, R. (2018, May 25). *How to Add and Subtract Positive and Negative Numbers*. Retrieved May 22, 2019, from http://www.mathsisfun.com/positive-negative-integers.html

Quinones, T. (2012, April 6). *SketchDaily - April 2012*. Retrieved from Flickr: https://www.flickr.com/photos/tomascosauce/7052032775/in/album-72157629369081946/

Rogers, L. (2008, January). *The History of Negative Numbers*. Retrieved from NRICH: https://nrich.maths.org/5961

Smith, B. S. (2015, Nov 15). *Fundamental Theorem of Arithmetic*. Retrieved May 25, 2019, from Platonic Realms Interactive Mathematics Encyclopedia:: http://platonicrealms.com/encyclopedia/Fundamental-Theorem-of-Arithmetic/

Wikipedia Contributors. (2019, April 30). *Divisibility rule*. Retrieved from https://en.wikipedia.org/w/index.php?title=Divisibility_rule&oldid=894867975

Wikipedia Contributors. (2019a, May 12). *Irrational number*. Retrieved May 18, 2019, from Wikipedia: https://en.wikipedia.org/w/index.php?title=Irrational_number&oldid=896684725

Wikipedia Contributors. (2019b, May 16). *Number*. Retrieved from Wikipedia: https://en.wikipedia.org/w/index.php?title=Number&oldid=897335076

Wikipedia Contributors. (2019c, May 15). *Order of operations*. Retrieved from Wikipedia: https://en.wikipedia.org/w/index.php?title=Order_of_operations&oldid=897269483

Wikipedia Contributors. (2019e, May 16). *Mill (currency)*. Retrieved May 26, 2016, from Wikipedia, The Free Encyclopedia.: https://en.wikipedia.org/w/index.php?title=Mill_(currency)&oldid=897300723

Wikipedia Contributors. (2019f, May 1). *Pythagorean theorem*. Retrieved May 31, 2019, from https://en.wikipedia.org/w/index.php?title=Pythagorean_theorem&oldid=894945625

Wikipedia Contriutors. (2019d, April 15). *Venn diagram*. Retrieved May 25, 2019, from https://en.wikipedia.org/w/index.php?title=Venn_diagram&oldid=892567228

GRAPH PAPER

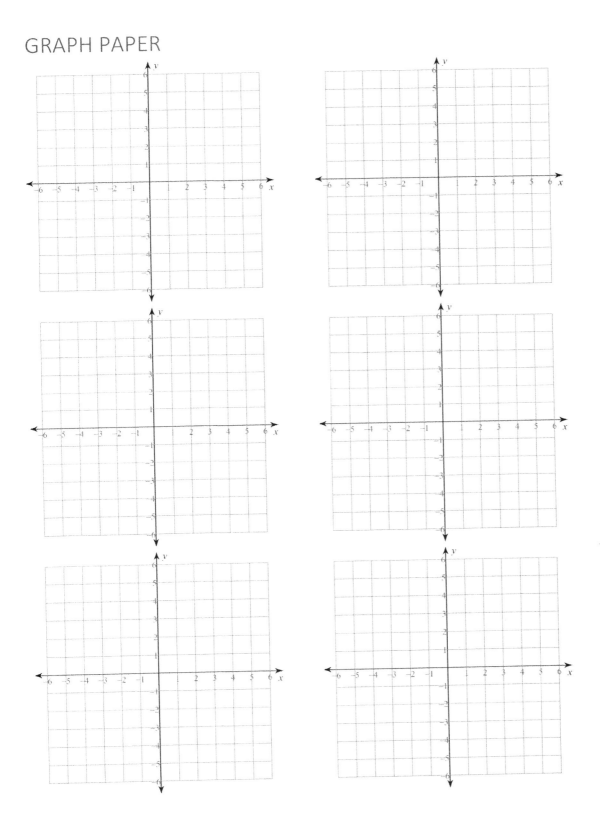

GRAPH PAPER

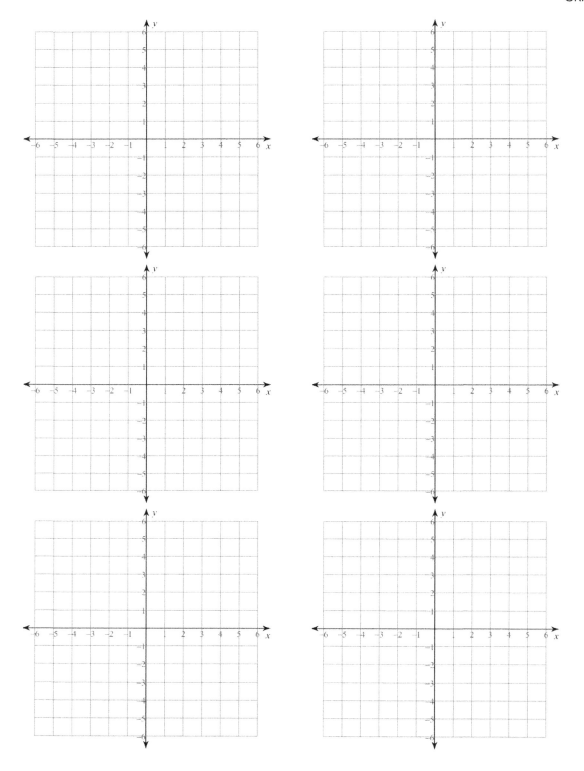

GRAPH PAPER

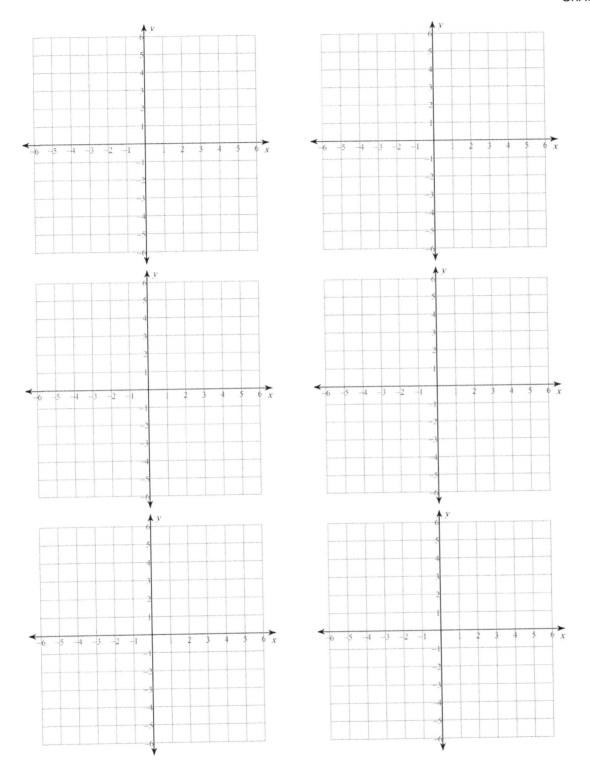

GRAPH PAPER

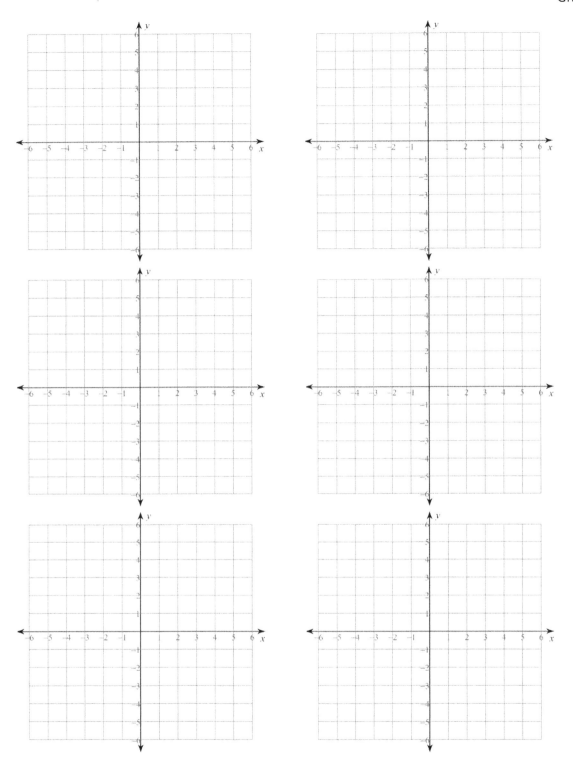

GRAPH PAPER

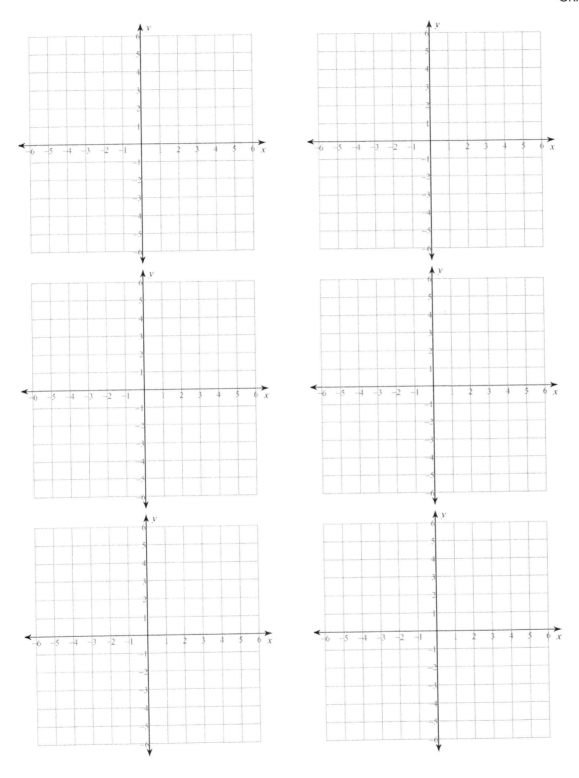

GRAPH PAPER

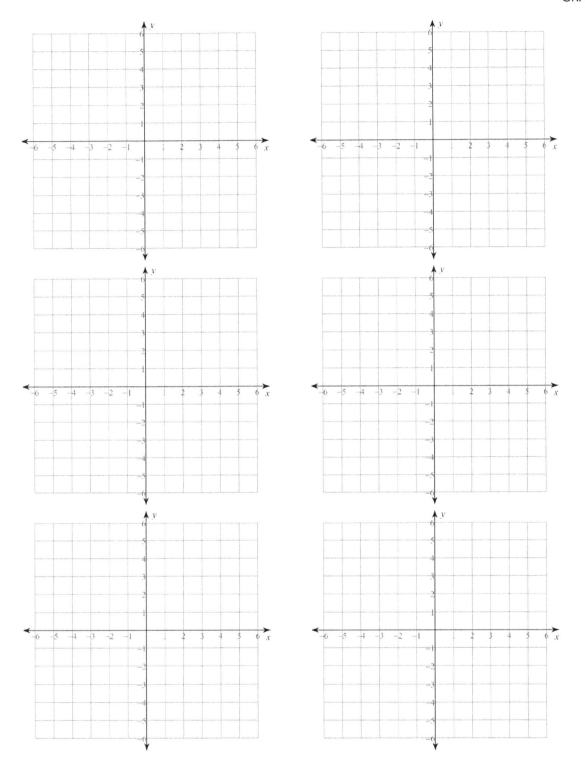

GRAPH PAPER

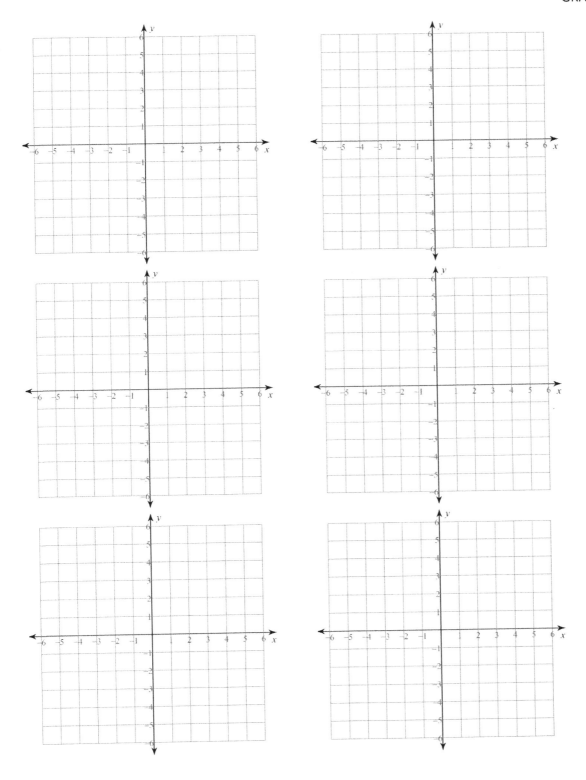

GRAPH PAPER

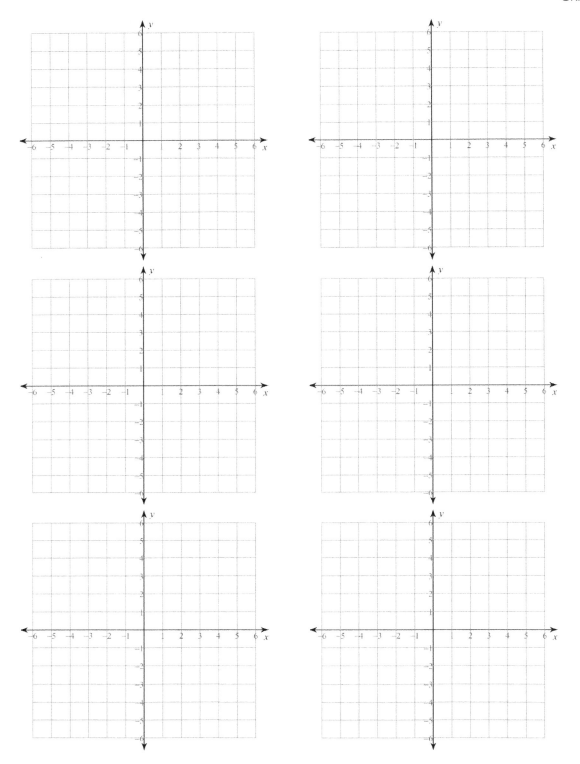

GRAPH PAPER

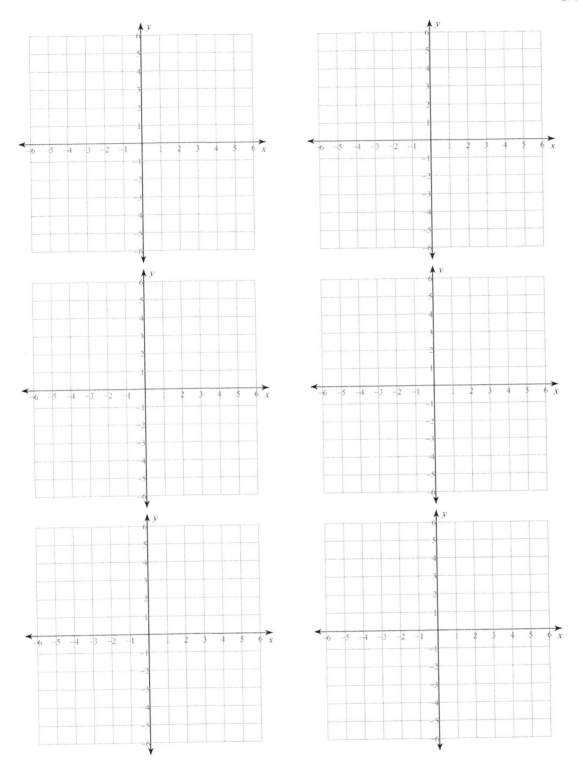

Page | 211

GRAPH PAPER

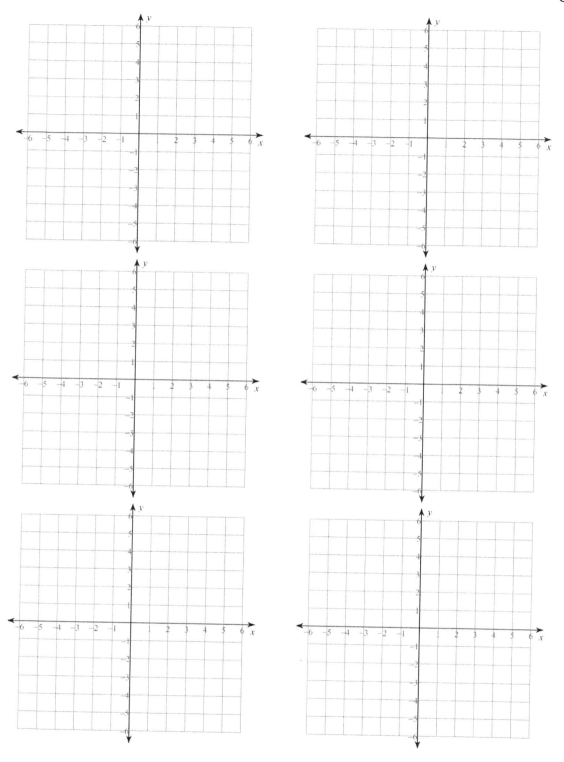

GRAPH PAPER

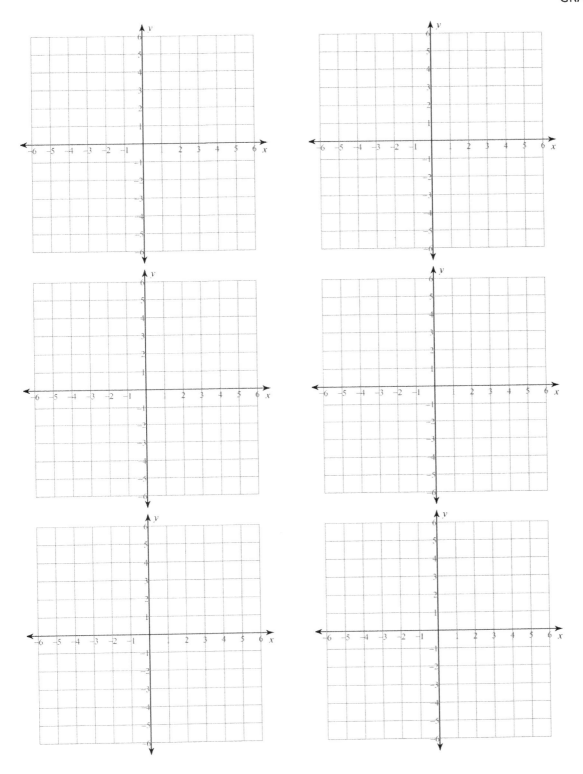

GRAPH PAPER

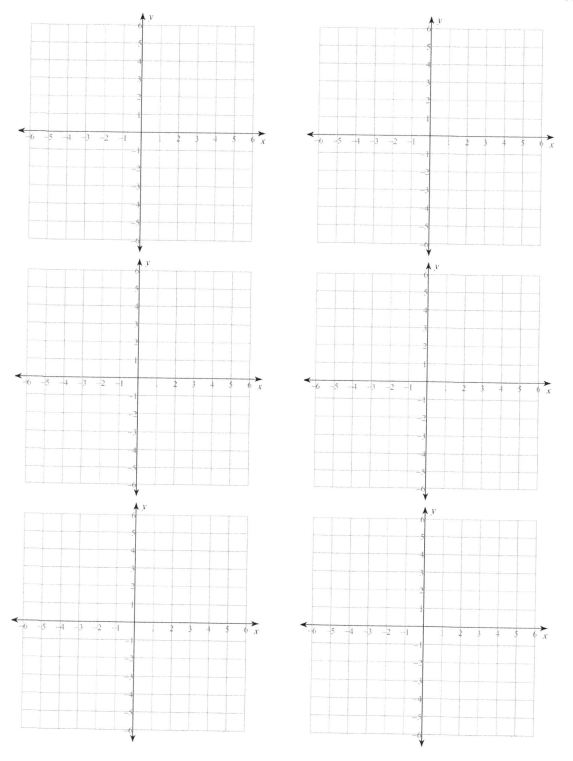

GRAPH PAPER

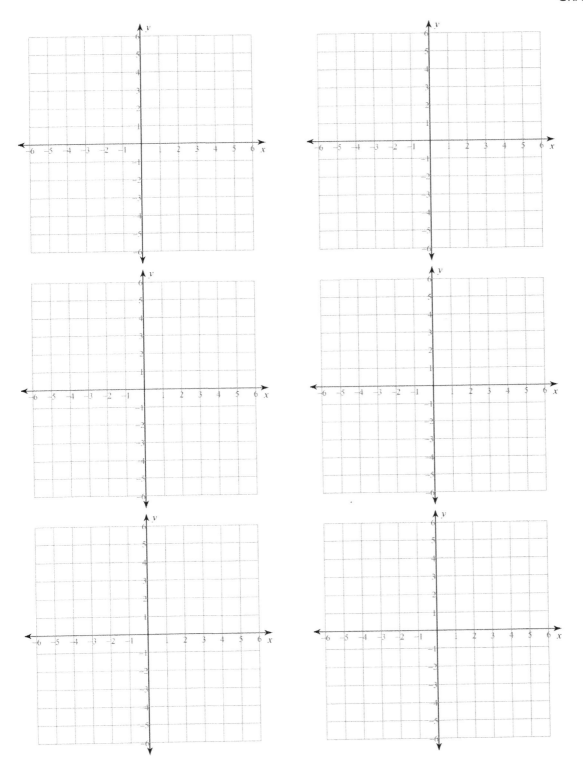

Page | 215

GRAPH PAPER

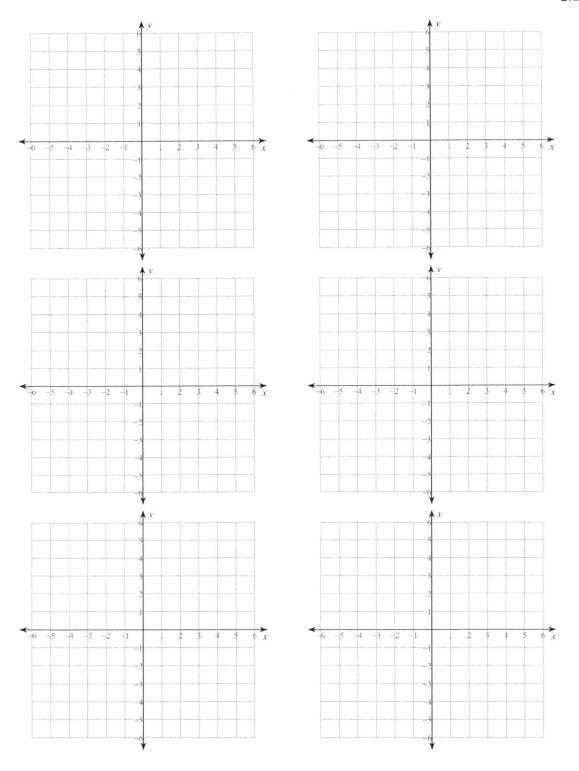

GRAPH PAPER

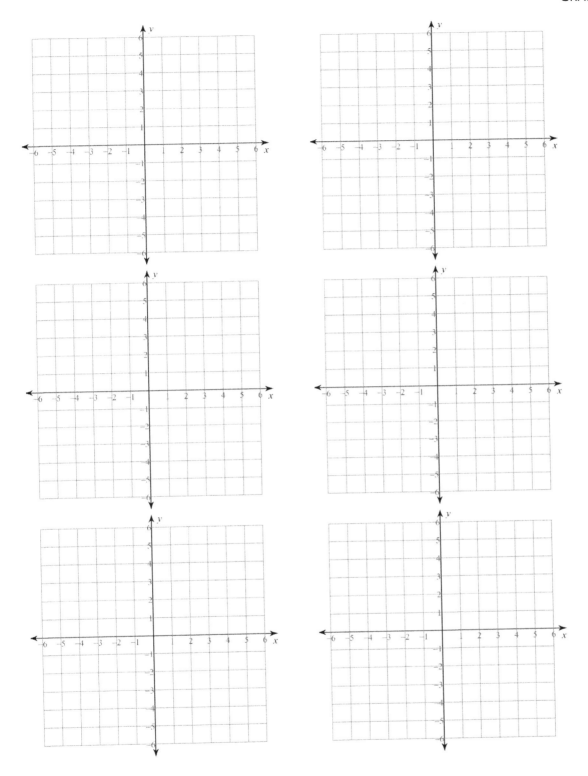

GRAPH PAPER

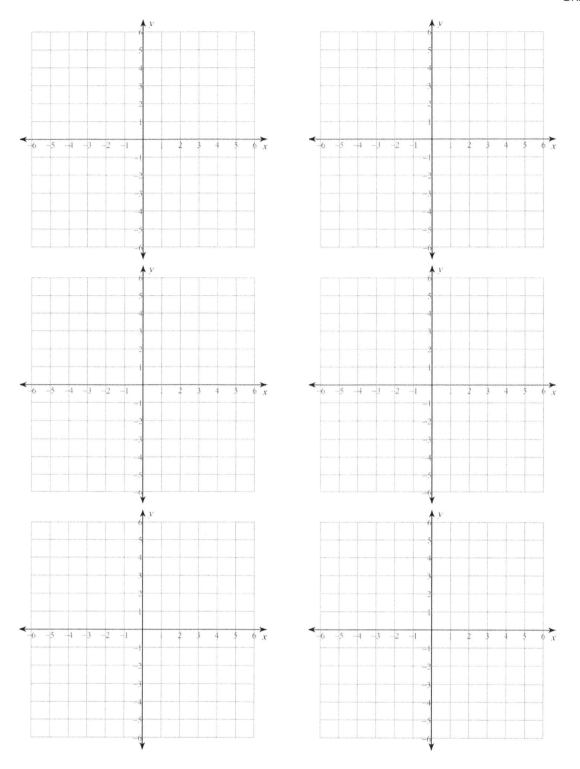

Page | 218

GRAPH PAPER

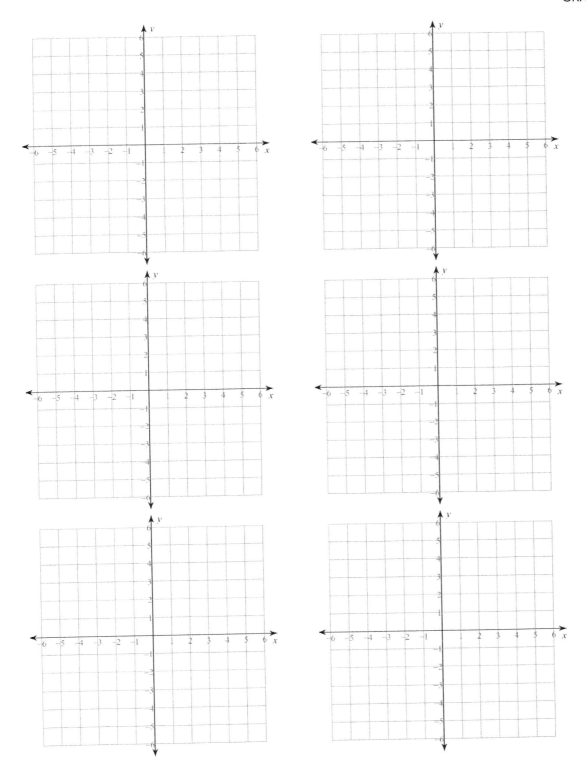

GRAPH PAPER

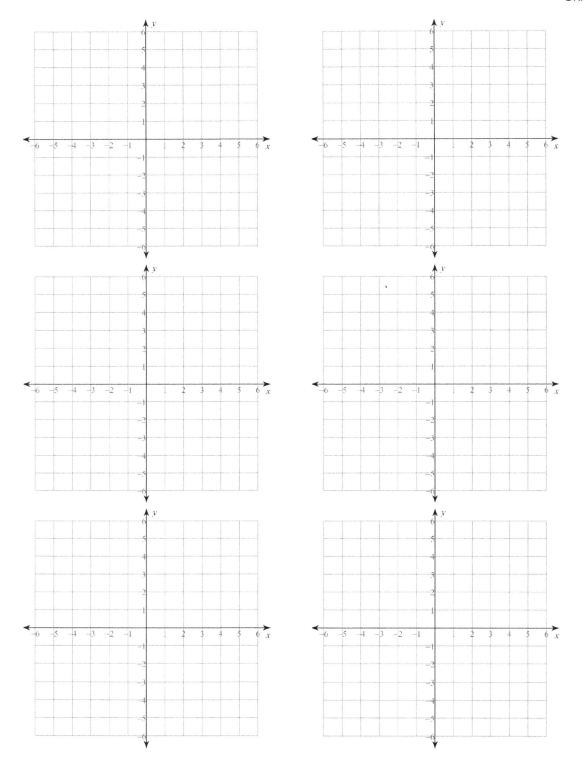

GRAPH PAPER

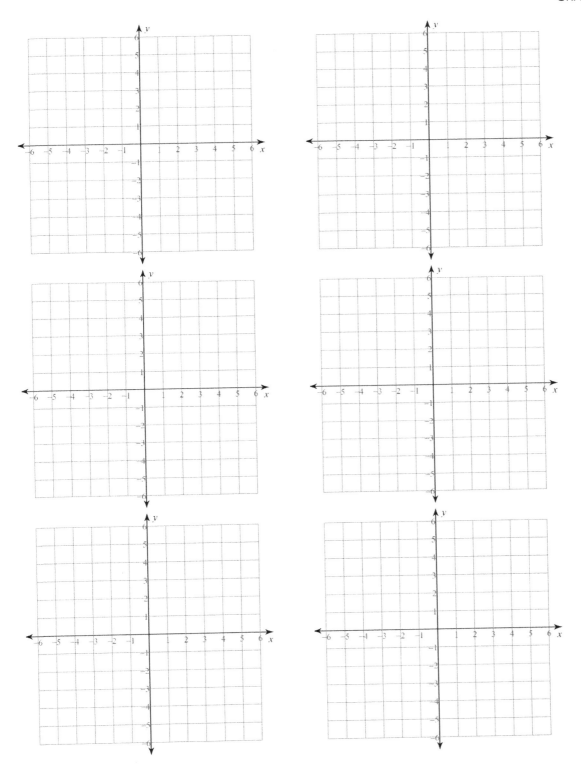

GRAPH PAPER

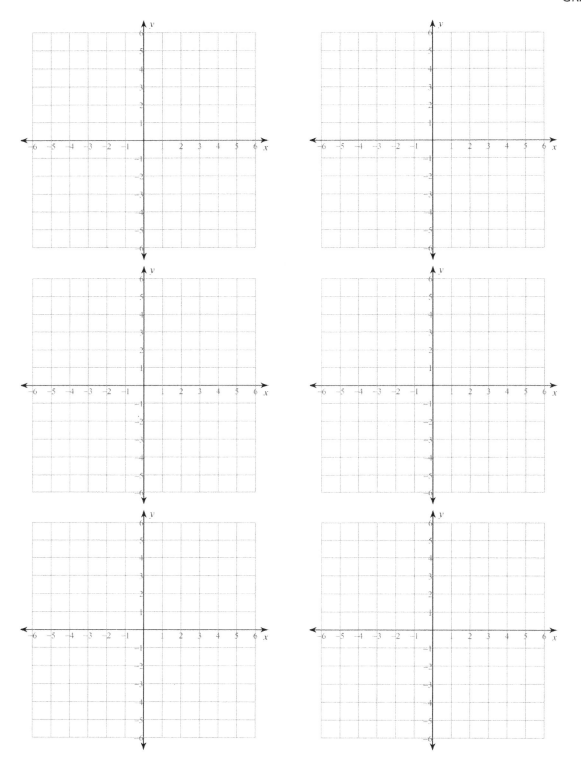

GRAPH PAPER

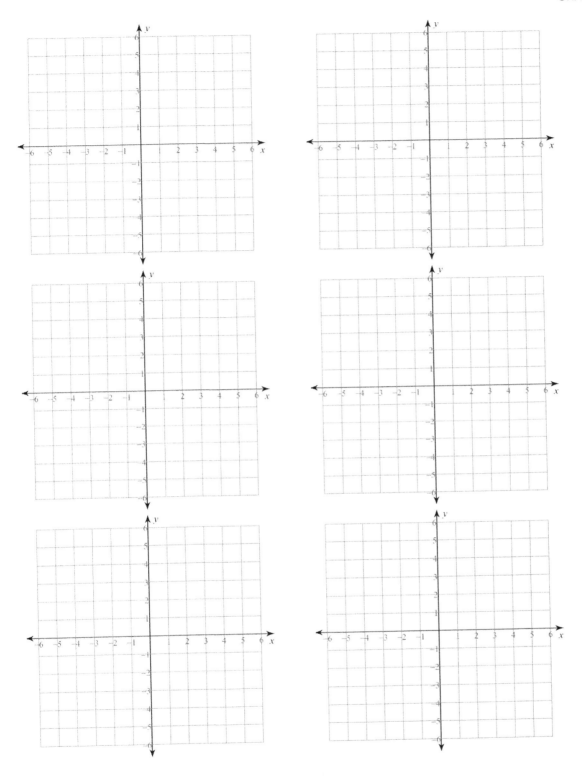

GRAPH PAPER

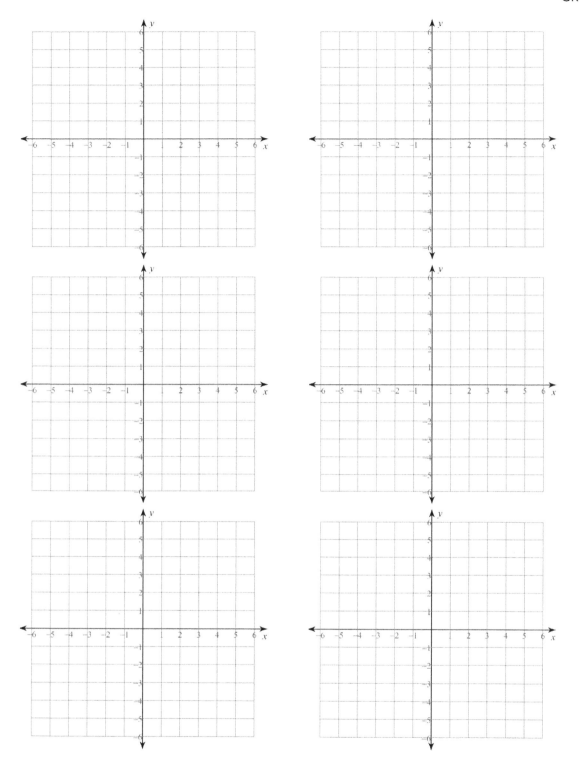

Page | 224

GRAPH PAPER

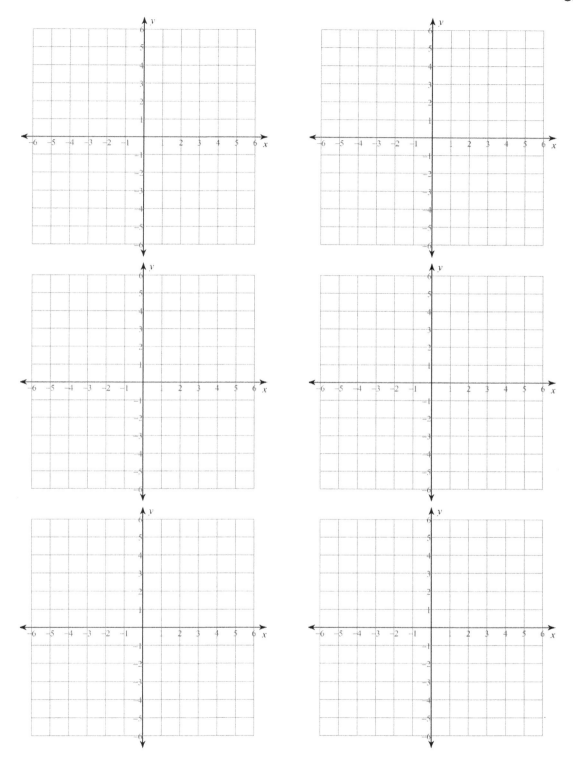

GRAPH PAPER

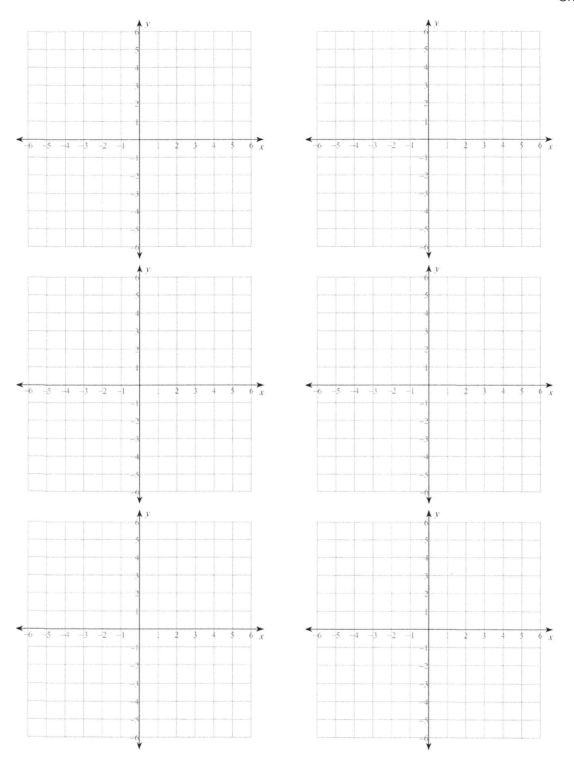

Page | 226

GRAPH PAPER

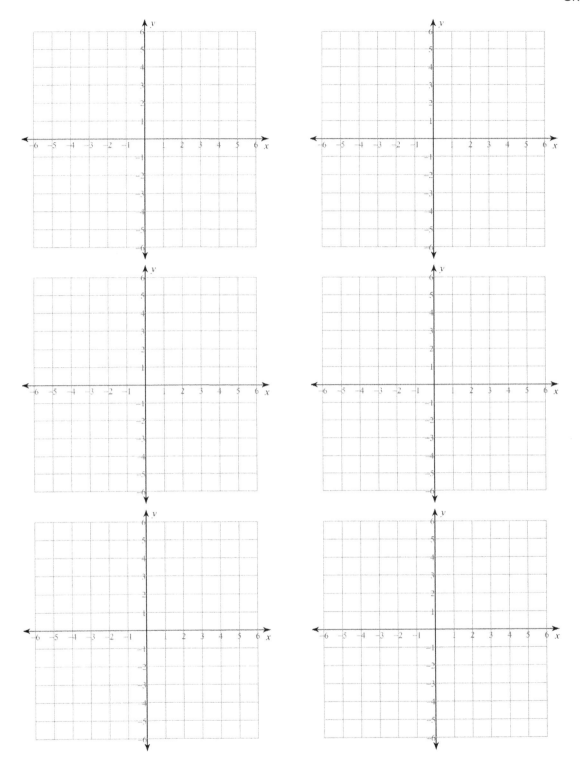

Page | 227

Made in the USA
Middletown, DE
16 March 2023

26945790R00126